Seismic Reinforcement of Infrastructure

Lessons from Catastrophic Earthquakes in Japan

Masanori Hamada

CRC Press
Taylor & Francis Group
Boca Raton London New York

CRC Press is an imprint of the
Taylor & Francis Group, an **informa** business

Designed cover image: iStock

First edition published 2027
by CRC Press
2385 NW Executive Center Drive, Suite 320, Boca Raton FL 33431

and by CRC Press
4 Park Square, Milton Park, Abingdon, Oxon, OX14 4RN

CRC Press is an imprint of Taylor & Francis Group, LLC

ISBN: 9781041219132 (hbk)
ISBN: 9781041217350 (pbk)
ISBN: 9781003728788 (ebk)

DOI: 10.1201/9781003728788

Typeset in Palatino
by codeMantra

Contents

Preface

Since completing my master's degree in 1968, I have been continuously involved in the research, development, and practical works in the field of earthquake disaster mitigation for over half a century. During this time, as an expert in earthquake disaster prevention, I experienced two major failures. The first failure was at the 1995 Kobe earthquake. Inland active faults near the highly urbanized area of the Osaka-Kobe region generated extremely intense ground motions, collapsing about 67,000 buildings and houses, resulting in the loss of over 5,500 lives.

The second failure was at the 2011 East Japan earthquake. A massive tsunami exceeding 10 m in height struck the coast along the Pacific Ocean from the Tohoku to the Kanto regions of the main island of Japan, resulting in over 25,000 deaths or missing persons. It was the worst disaster in Japan next to the 1923 Kanto earthquake, which killed more than 100,000 people. Notably, the Fukushima Daiichi Nuclear Power Plant experienced the meltdown of nuclear reactors, an unprecedented accident in world history. The plant's design was based on a tsunami height of about 6 m, but it was overwhelmed by a tsunami exceeding 15 m, resulting in the loss of all cooling functions. The failure to accurately predict the tsunami height for the design was the origin of the accident.

Seven years before the East Japan earthquake, in 2004, I visited Banda Aceh at the northern tip of Sumatra Island, Indonesia, which had lost 70,000 lives, about one-third of the total population of the city, by an earthquake of magnitude 9.1 and a consequent huge tsunami. Seeing the devastation where most buildings and houses had collapsed and were swept away to the sea by the tsunami, I thought, without any scientific basis, that such a disaster would not occur in Japan. Most Japanese experts, including myself, believed that earthquakes exceeding moment magnitude 9 would not happen in Japan and that concurrent Tokai and Tonankai earthquakes of magnitude 8.5 along the Nankai Trough in the Pacific Ocean would be the largest possible in our country.

In the seismic design of civil and architectural structures such as buildings and bridges, the magnitude of the earthquake, the ground motions for the design, and the height of the tsunami are assumed at a proper level by taking the history of past seismic activities, existing active faults, and geological conditions. However, the ground motions caused by the Kobe earthquake and the tsunami induced by the East Japan earthquake largely exceeded those assumed for the original design. Both disasters taught us that we must prepare strategies and develop measures for the prevention of catastrophic impacts on the nation and human lives, even when natural phenomena like earthquake, tsunami, and rainfall exceed the assumed level for the design.

This book describes the background of my research and development for earthquake disaster mitigation over half a century, including the author's societal activities and international cooperations.

A focus of my research in later years was the "liquefaction-induced large ground displacements." Triggered by investigating damage to gas pipelines by the 1983 Japan Sea earthquake, I focused on this phenomenon, researching the damage caused by the ground displacements and the methods to predict ground displacements, and finally, I developed a countermeasure, "skip-driven pile method," to protect industrial facilities against ground displacements in reclaimed coastal areas, which has been realized for a national project "Enhancement of National Land Resilience." It was fortunate that my research and development activities encompassed the entire process, from discovering the phenomenon, clarifying its mechanism, and proposing ground displacement prediction methods to developing countermeasure techniques.

Japanese seismologists have warned of the imminent occurrence of an earthquake directly underneath Tokyo, the capital of Japan, and a major earthquake along the Nankai Trough in the Pacific Ocean. Predicting the time within several years of occurrence is difficult, but these earthquakes will inevitably happen over a long period.

I hope this book will provide valuable insights and information for researchers and practitioners in the field of earthquake disaster engineering, contributing to future research and development for disaster mitigation and national land resilience.

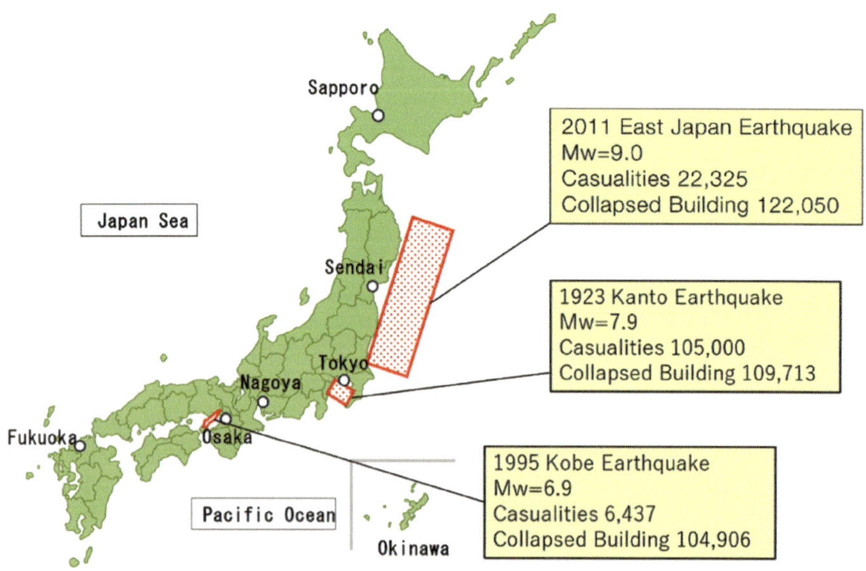

Catastrophic earthquakes during last century in Japan

About the Author

Masanori Hamada, Professor Emeritus, Waseda University, is internationally recognised as an authority for the analysis and mitigation of earthquake and tsunami disasters. He was engaged in the earthquake-resistant design and the construction of bridges, dams, port and harbor structures, and nuclear power plants. He was the 43rd president of the Japan Society of Civil Engineers and was elected as an international member of the National Academy of Engineering, USA.

1

Lessons Learned from Past Catastrophic Earthquakes in Japan

1.1 The 1923 Kanto Earthquake[1]

-The beginning of earthquake engineering in Japan-

One century has passed since the 1923 Kanto earthquake of moment magnitude (Mw) 7.9. This earthquake, with over 100,000 casualties, and 570,000 buildings and houses collapsed or burned, stands as the worst natural disaster in Japan.

Buildings constructed with modern technologies from the West since the Meiji Restoration in 1868 also suffered considerable damage. Those buildings had been designed without considering any earthquake effects. The devastation to buildings by the earthquake and its consequent fire led to the beginning of seismic resistance designs, in which statical inertial forces induced by an earthquake acceleration are taken into consideration, besides the structure's weight as an external force, to evaluate the structure's stability, deformation, and stresses.

The problem was how much statical force should be used for the design. The inertial force is the product of the structure's masses and the earthquake-induced accelerations. At the time, there was insufficient acceleration data, leaving not enough data basis to evaluate the inertia forces from earthquake acceleration. As a practical method, certain coefficients, horizontal seismic coefficient K_H and vertical seismic coefficient K_V, were multiplied by the structure's weight to determine the static design force. These coefficients are known as the "design seismic intensity," different from the "Seismic Intensity Scale" released by the Japan Meteorological Agency, and Modified Mercalli Intensity.

Initially, many structures were assigned a design seismic intensity K_H of 0.1, meaning the design accounted for 10% of the weight acting horizontally. Later, considerations were made for the structure's importance and the potential risk of being destroyed by an earthquake. Nowadays, a design seismic intensity of 0.2 is commonly used for most infrastructures. For nuclear facilities and high-pressure gas facilities, a high coefficient of 0.6 or more is used.

When converted to acceleration, a design seismic intensity of 0.2 corresponds to about 200 cm/s^2 multiplied by the gravitational acceleration of 980 cm/s^2.

DOI: 10.1201/9781003728788-1

Surface accelerations during the Kanto earthquake were recorded at several points in the Tokyo area at the time. These records suggest maximum surface accelerations of 200–300 cm/s^2, indicating that the design seismic intensity of 0.2 was based on the lower value of earthquake shaking observed in Tokyo during the Kanto earthquake.

This design seismic intensity has been a standard in seismic design since the Kanto earthquake. However, the 1995 Kobe earthquake recorded exceedingly high accelerations, over 800 cm/s^2 in Kobe, significantly surpassing the accelerations anticipated in seismic designs since the Kanto earthquake.[2,3]

The Kanto earthquake, with its epicenter in Sagami Bay along the Pacific Ocean and a magnitude of 7.9, had an epicentral distance of about 100 km from Tokyo. In contrast, the Kobe earthquake, with a magnitude of 7.3, was caused by an inland fault. The earthquake, occurring near the active fault, was less than 10 km to the Osaka-Kobe highly urbanized area, generated large accelerations, which were not accounted for in the seismic design before the Kobe earthquake.

The extensive damage to buildings, bridges, and lifeline facilities by the Kobe earthquake led to consideration of earthquake motions, like those observed in the Kobe area, as Level 2 earthquake motions in addition to the conventional Level 1 earthquake motions used in seismic design since the Kanto earthquake (Figures 1.1.1 and 1.1.2).

FIGURE 1.1.1
Catastrophes by the 1923 Kanto earthquake and international aid. (Quoted from the home page of the National Museum.)

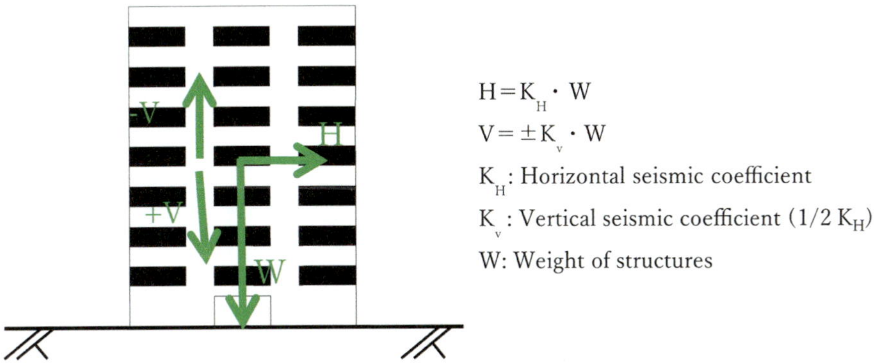

FIGURE 1.1.2
Earthquake-resistant design by the seismic coefficient method.

1.2 The 1995 Kobe Earthquake[4]

1.2.1 Day of the Earthquake

The day before the Kobe earthquake of moment magnitude (Mw) 6.9 occurred on January 17, 1995, on a Sunday evening, my wife and I had dinner near Shin-Yokohama Station and then took the Shinkansen around 7 p.m. heading for Osaka. This was to attend the "First Japan-US Urban Disaster Prevention Conference" starting the next day, where I was designated as the chairman for the opening session at 9 a.m. I checked into a hotel near Osaka Castle around 11 p.m. and went to bed early to prepare for the conference beginning the next day.

At 4:45 a.m. the following morning, I was almost thrown out of bed by an intense tremor unlike any I had experienced before in the Tokyo and Yokohama area. I immediately feared that the anticipated Kanto earthquake had struck the Tokyo and Yokohama area. I was concerned about the situation in Yokohama, where my home was, but the phone in the room was dead and unusable.

Meanwhile, a television broadcast started, and the first image showed smoke rising from multiple points in the urban area, suggesting fires caused by the earthquake. Figure 1.2.1 was aerial shots taken by the Japan Air Self-Defense Force, indicating that fires were caused after the earthquake at several locations in Kobe.

At that time, most seismologists of Japan did not predict a major earthquake occurring in the Osaka and Kobe areas. As a researcher in earthquake engineering myself, I found it hard to believe that the epicenter was in Kobe.

The hotel I stayed at, which was about 50 km away from the epicenter, suffered minor damage, such as part of the exterior wall collapsing, but the building structure itself showed no significant damage. After having a simple

FIGURE 1.2.1
Earthquake-consequent fire (1995 Kobe earthquake).

breakfast prepared by the hotel, I headed to the venue of the disaster prevention conference. Several attendees had already gathered and were engrossed in the television news, which indicated that many buildings and houses had collapsed, and fires were breaking out at multiple locations. About 50 people from the United States were expected to attend the conference, and more than half of them had arrived at the venue, sharing their experiences of the severe tremors. Although earthquakes occasionally occur on the West Coast of the United States, with intensities around Level III on the Japan Meteorological Agency (JMA) Scale, the Rank VI intensity experienced this time was a first for the attendees from the United States. I had experienced up to Rank V before, but this was also my first experience with tremors above Rank VI.

The decision was made by the participants at the conference to proceed with the opening session regardless. All formal speeches were cancelled, and discussions focused on a plan for a joint investigation between Japan and the United States for the earthquake damage. Groups were formed with each focusing on a specific aspect of the investigations, with plans to reconvene in the evening to share information.

I decided to investigate the damage in the artificial islands reclaimed from the sea in Kobe with Professor S from Kyoto University. We rented a car and headed to Kobe. The first site we encountered was an elevated highway at Fukae ward in Kobe, where the bridge piers had continuously collapsed over a distance of more than several hundred meters (Figure 1.2.2a). We were just stunned to see such damage occurring in our country.

(a) Highway bridge piers collapsed continuously

(b) Concrete piers of shinkansen (the bullet train) damaged by shear force

(c) The columns of this subway were ruptured by shear force

FIGURE 1.2.2
Collapse of concrete piers of highway, railway, and subway.

Approximately one year before the Kobe earthquake, the Northridge earthquake struck California, USA, in January 1994, causing the collapse of freeways. A Japanese research team was sent to the United States, and in response to U.S. media inquiries, a leading researcher stated, "Japan's bridge seismic design is stricter than that of the United States, and the seismic forces used in the design are also larger than those in U.S. Such damage that occurred in California should not occur in Japan." I had thought the same. However, about only one year later, damage exceeding that in California occurred in Japan. The main cause was the strong seismic motions generated by close inland faults. The seismic motion induced by a magnitude 7 class inland fault was not considered in the design of structures before the Kobe earthquake.

The seismic design of structures in Japan started after the 1923 Kanto earthquake, based on the tremors observed in the Tokyo metropolitan area over 50 km away from the epicenter in the Sagami Bay. In contrast, the distance from the inland fault that caused the Kobe earthquake to urban areas like Kobe was less than 10 km. The magnitude 7.3 of the Kobe earthquake was smaller than the magnitude 7.9 of the 1923 Kanto earthquake, but highly intense ground motion was generated near the fault.

After the Kobe earthquake, the principles of seismic design codes for infrastructures and lifeline facilities were revised, and the seismic forces for the design were significantly increased. This will be described in Section 1.2.3.

When we arrived at the site of the collapsed elevated highway bridges, the media had already arrived, and several reporters surrounded us, asking for our thoughts on the damage and seeking expert opinions. Professor S and I were very much confused by the extent of the damage and could not provide any comment.

Professor S had to go to another site with a different media team, so I drove the rental car alone toward the center of Kobe City. As the day began to darken when I reached Sannomiya, downtown of Kobe, I realized I had to return to Osaka. I started to head back along the collapsed national highway, but encountered a massive traffic jam. Unable to move forward, I took a side road, only to find it completely blocked by debris from collapsed houses. I left Osaka without having lunch and brought neither drinking water nor emergency food to Kobe. Trapped in the car, I was unable to grasp the full extent of the disaster. The radio broadcasts alone could not provide a clear picture of the earthquake damage. Fortunately, I had an early model mobile phone, which I used to contact my wife at home in Yokohama and get an update on the disaster situation from television news reports.

The car moved slowly, and it was after 4 a.m. the next day, about 16 hours after leaving Osaka, when I finally entered the city. It was still before dawn, and Osaka seemed as if nothing had happened, completely silent. There was no visible earthquake damage.

I first tried to return to the hotel I had stayed in the day before. I remembered the name of the hotel but I was not familiar with Osaka's geography. I caught a taxi and asked the driver to take me to the hotel by leading my car, and I finally arrived there after much difficulties. By then, almost one day had passed since the earthquake struck.

1.2.2 Lessons Learned from the Earthquake

The Kobe earthquake was caused by an inland fault with a moment magnitude of 7.3. The fault stretched for 30 km from Awaji Island to the Rokko Mountain Range in Kobe (Figure 1.2.3), and because the epicenter was shallow at 10 km, it generated intense ground motions in the urban areas of the Kobe-Osaka region.

The horizontal acceleration at the surface recorded by the Kobe Marine Meteorological Observatory was reported to be about 800 cm/s^2. The seismic intensity level by the JMA scale reached the highest rank VII.

Earthquakes by inland faults with magnitudes over 7, such as the 1891 Nobi earthquake of magnitude 8.0 and the 1948 Fukui earthquake of magnitude 7.1, have occurred approximately once every decade in Japan. A report from the Fukui earthquake describes bamboo groves being uprooted and thrown into the air, hinting at the strength of the earthquake's tremors, particularly in the vertical direction. However, magnitude 7-class inland earthquakes in Japan have been considered less likely to hit major urban areas directly compared to magnitude 8-class earthquakes occurring at plate boundaries along

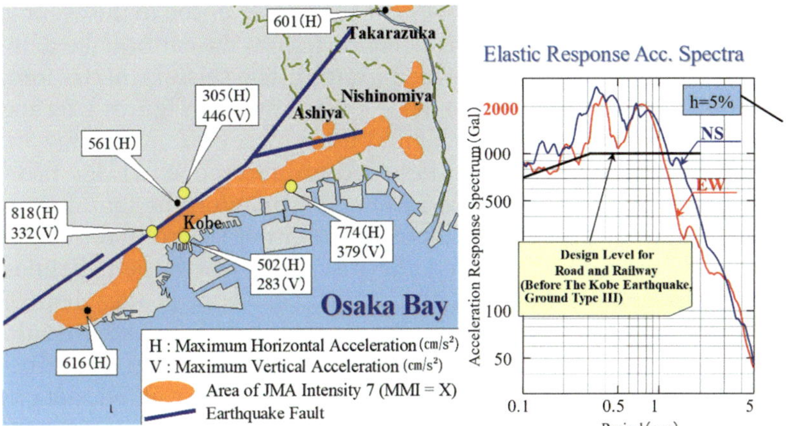

The maximum acceleration 818 cm/s² was recorded at Kobe Marine Meteorological Observatory

(a)

Acceleration Response Spectrum reached over 2000 cm/s²

(b)

FIGURE 1.2.3
Earthquake faults and ground motions: (a) acceleration recorded on ground surface and (b) response spectra.

the Pacific Ocean, because the area affected by strong shaking is limited in the case of inland fault.

The intense tremors that struck Kobe and Osaka urban areas caused the collapse of over 100,000 buildings and houses, claiming the lives of more than 5,500 people. Including those who died related to the earthquake aftermath, the death toll exceeded 6,600, making it one of the deadliest disasters in Japan since the 1923 Kanto earthquake.

As mentioned in Section 1.1, the design inertial force, evaluated in terms of acceleration, was about 200–300 cm/s^2 before the Kobe earthquake, significantly less than the 800 cm/s^2 observed in Kobe.

The most important lesson from the Kobe earthquake regarding the seismic design of structures was that structures should be designed to prevent total collapse, even when subjected to seismic motions exceeding the assumed levels. The fundamental principle is that if a structure does not collapse despite significant damage and large deformation, it can prevent loss of life. This requires structures to have not only strength but also ductility. More details of this principle are introduced in the Section 1.2.3.

Damage to the concrete piers of Shinkansen (the bullet train, Figure 1.2.2b) raised concerns among both civil engineers and the public about the seismic safety of high-speed railway structures. Concrete piers were destroyed by horizontal forces (a type of failure known as shear failure), causing the bridge girders to fall. The severely damaged piers lacked sufficient ductility. Fortunately, the earthquake occurred before the start of railway service operations, avoiding a major disaster with numerous casualties, but it highlighted the issue of safety for high-speed railways like the Shinkansen in the event of an inland fault earthquake. A significant concern from the earthquake damage was lateral displacements of liquefied ground. The seismic inertia force and liquefaction caused quay walls to move toward the sea (Figures 1.2.4 and 1.2.5), displacing the reclaimed land behind them by several meters toward the sea. Horizontal displacements of liquefied ground moved bridge piers, causing bridge girders to fall (Figure 1.2.6) and damaging the foundations of buildings and structures. Furthermore, the strain of the ground (relative displacements of the ground) caused the disruptions of buried lifelines. Horizontal displacement of liquefied ground had been a focus after the 1983 Japan Sea earthquake, but the earthquake occurred before research results could be reflected in seismic design and reinforcement, leading to recurrence of the same damage. After the earthquake, the seismic design of foundations of structures and buried structures like lifeline pipes began to consider the effects of horizontal ground displacements.

When reflecting on the disaster lessons learned from the Kobe earthquake, it is essential to remember the disasters that had not occurred under specific conditions. The Shinkansen is one such example. If the earthquake had occurred just one hour later, derailment of the running Shinkansen would have been inevitable. Trains running on elevated tracks could have jumped the tracks and fallen onto houses and buildings below, potentially leading to an even greater disaster. The same can be said for the collapse of subway stations. Concrete ceilings fell onto the tracks along with the soil above them

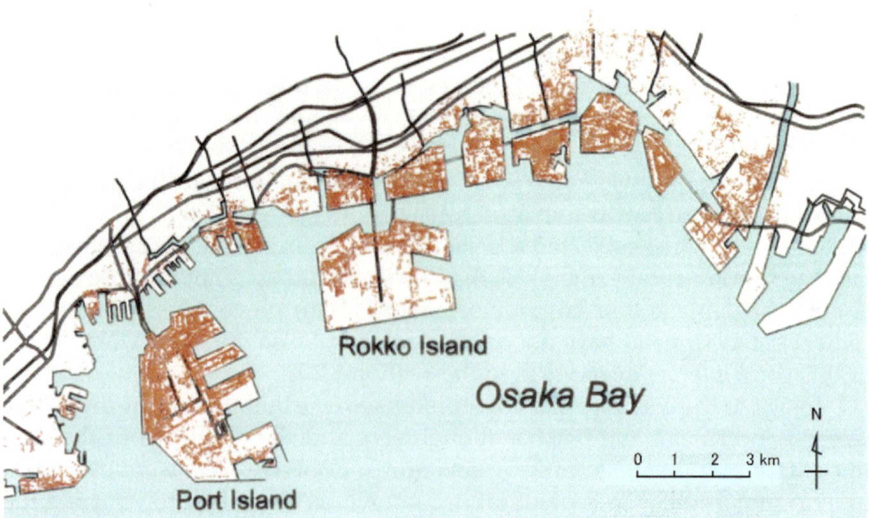

FIGURE 1.2.4
Liquefaction in man-made islands reclaimed from the Osaka Bay (1995 Kobe earthquake).

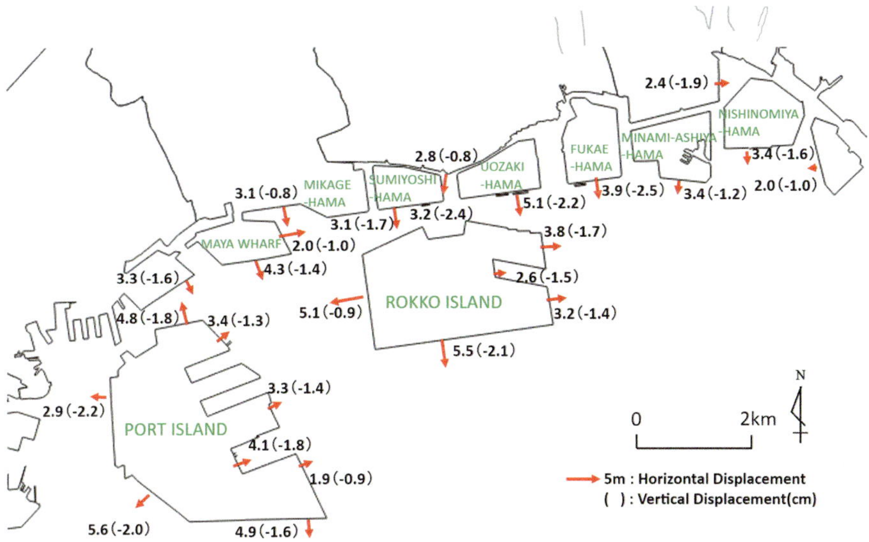

FIGURE 1.2.5
Quay wall displacement due to soil liquefaction (1995 Kobe earthquake).

FIGURE 1.2.6
Fallen girder of a highway bridge due to liquefaction-induced ground displacement (1995 Kobe earthquake).

(Figure 1.2.2c). If a train had been stopped there or collided with the collapsed section, the disaster would have escalated.

Disasters that did not occur under certain conditions can be cited in both hardware and software aspects. One such factor was the early morning timing of the earthquake. If the quake had occurred 2–3 hours later, coinciding with the morning rush hour, the consequences could have been even more severe. Conversely, the fact that it was dawn after the earthquake helped with emergency and evacuation efforts for people trapped under collapsed buildings. If the earthquake had occurred at midnight, the number of casualties might have been higher.

1.2.3 New Seismic Design

Following the Kobe earthquake, the Japan Society of Civil Engineers (JSCE) organized a special committee on earthquake resistance enhancement of civil engineering structures, composed of expert engineers and researchers from various fields of civil engineering. As a result, the following two basic policies were proposed to enhance the seismic performance of civil engineering structures (Figure 1.2.7):

i. Seismic design against two-level ground motions
ii. Performance-based design

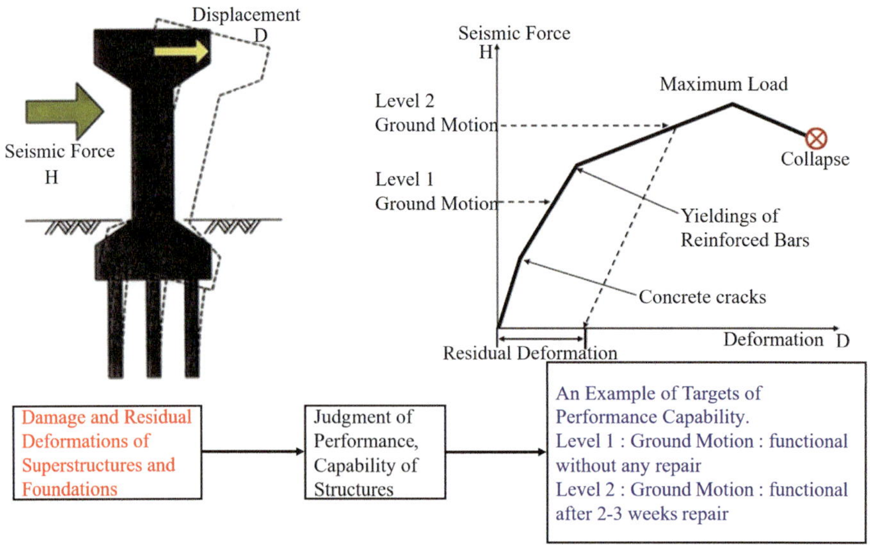

FIGURE 1.2.7
Performance-based seismic design against two-level ground motions (example of railway bridges).

The two-level earthquake motions design approach adds to the conventional seismic design considerations by including not only the design seismic motion (Level 1 seismic motion) traditionally considered but also an examination of seismic resilience against more intense ground motions such as those observed during the Kobe earthquake. Level 1 seismic motion is based on the tremors experienced in Tokyo during the 1923 Kanto earthquake, with accelerations considered in the range of 200–300-gal cm/s^2. In contrast, Level 2 seismic motion assumes accelerations 800-gal cm/s^2 and more, such as observed on the ground surface in Kobe.

Performance-based design specifies the performance (functionality) that structures should maintain during and after an earthquake when subjected to the design ground motions. This design approach aims to allow the extent of structural damage and residual deformation within acceptable limits to ensure the required performance during and after the design earthquake. This method not only defines the performance of individual structures and facilities but also considers the performance of systems composed of structures and facilities, such as lifeline systems like water and electricity, by setting post-earthquake capabilities of the components and determining the functionality of the system.

The recommendations by JSCE for two-level seismic design and performance-based design were incorporated into the National Disaster Prevention Basic Plan by the Japanese Government. Following the Kobe earthquake, seismic design standards for most civil engineering structures were revised in accordance with these recommendations. However, several challenges were encountered during these revisions.

The first subject was how to determine the Level 2 seismic motion. Two approaches were proposed: one based on the "historical maximum" concept, which adopts the upper limit of observed ground motions to date as future design motions. This is why the maximum surface design seismic coefficient in revised standards after the Kobe earthquake was adopted as 0.8, based on the 800 cm/s^2 observed on the ground surface at the Kobe Marine Meteorological Observatory.

The other approach is to define the design of ground motion probabilistically, which was adopted for seismic standards in railway and water supply facilities after the earthquake. The "historical maximum" approach requires revising seismic motions every time a motion exceeding previous records is observed. On the contrary, the probabilistic approach needs accumulations of sufficient data to define the probability of extremely rare (low probability) ground motions, making it difficult to determine which approach is superior.

In performance-based design, the functionality a structure retains is judged based on the extent of damage and residual deformations of structural components, including the foundations. This requires a quantitative evaluation of the deformation process after the yielding (where residual deformation occurs) to failure. Traditional seismic design of structures aimed to prevent residual deformation post-earthquake, i.e., design within the elastic region before yielding. Therefore, research on the behavior and failure processes of steel and concrete structures in the plastic region had been conducted through a national research project after the Kobe earthquake.

Performance-based design was also applied to structures made primarily of soil, such as earth dams, embankments, quay walls, and retaining walls. Traditional designs of these structures verified that the resistance against sliding, overturning, etc., due to seismic horizontal forces and self-weight, increased or decreased by a vertical seismic force. Considering Level 2 ground motion in seismic design often results in external forces exceeding resistance, leading to sliding, tilting, subsidence, and horizontal displacement of structures, which are then checked for functionality based on these damages. However, reliable methods to estimate the extent of damage and residual deformation of soil structures like embankments and retaining walls have yet to be established.

Although many investigations and studies have been conducted on the process leading to structural failure after the Kobe earthquake, numerous issues still need to be resolved. The revision of seismic design standards after the Kobe earthquake was hastily implemented for the design of new reconstructions and repair of the damaged structures. Future revisions of seismic standards will need to consider ongoing research findings.

The two-level ground motion design and performance-based design concepts advocated by JSCE have been reflected in the seismic design of lifeline facilities such as gas, water, electricity, and telecommunications.

1.3 The 2011 East Japan Earthquake[5]

1.3.1 Failure of Earthquake and Tsunami Prediction

On March 11, 2011, when the East Japan earthquake of moment magnitude (Mw) 9.0 occurred, I was in my lab on the eighth floor of the ten-story school building at Waseda University, Tokyo. The intense shaking caused piles of documents and books on my desk to collapse and scatter across the floor. The bookshelf also swayed significantly, and I had to brace it with my body to prevent it from falling over. Once the shaking subsided, I looked outside the window and saw that the buildings in the neighborhood of the campus seemed to have suffered no noticeable damage. The 18-story main building of the school also appeared to be intact.

Shortly after, radio news reports on TV indicated that the earthquake source area was off the coast of the Tohoku region (northeast of the main island of Japan) in the Pacific Ocean as shown in Figure 1.3.1, with a

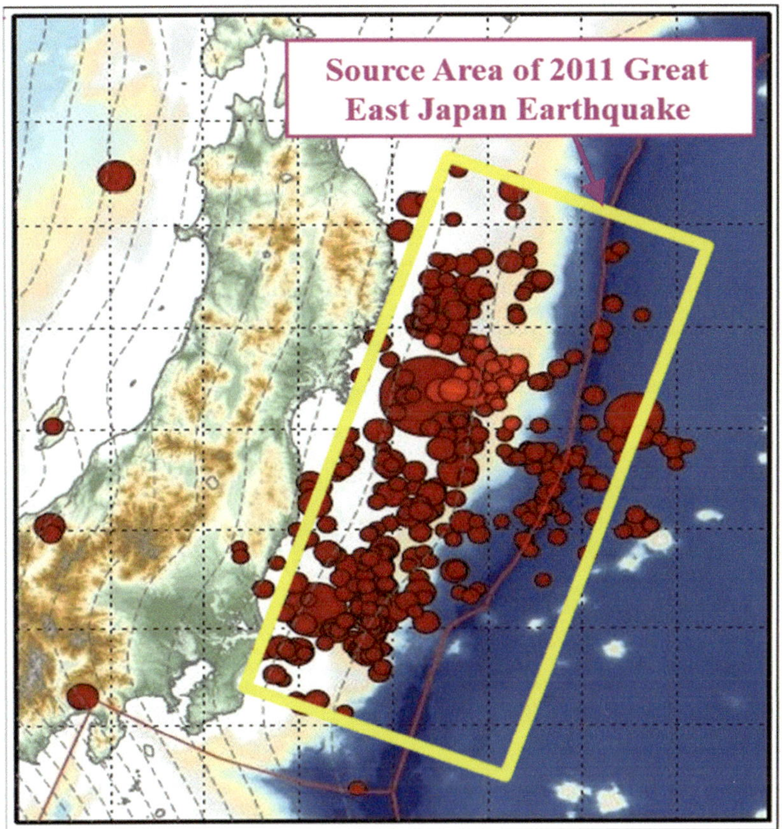

FIGURE 1.3.1
Source area of the 2011 East Japan earthquake and epicenters of main and aftershocks.

magnitude of over 8. The magnitude of the earthquake was revised several times afterward, finally settling at 9.0. Most experts in earthquake prediction, including those at the Japan Meteorological Agency, had previously believed that the total magnitude would be around 8.5 at maximum, even when the two earthquakes at Tokai and Tonankai along the Nankai Sea Trough in the Pacific Ocean, occur continuously, as shown in Figure 1.3.2. When a magnitude 9.1 earthquake struck off the coast of Sumatra, Indonesia, in 2004, the belief that Japan would not experience an earthquake exceeding magnitude 9 was accepted without question.

The Meteorological Agency and the Central Disaster Prevention Council of Japan reported that even if the Tokai and Tonankai earthquakes along the Nankai Sea Trough co-occurred, their combined magnitude would not reach 9.0.

The plate structures along the coast of Sumatra and the Pacific coast of the Tohoku region in Japan are similar as shown in Figure 1.3.3. The Japanese

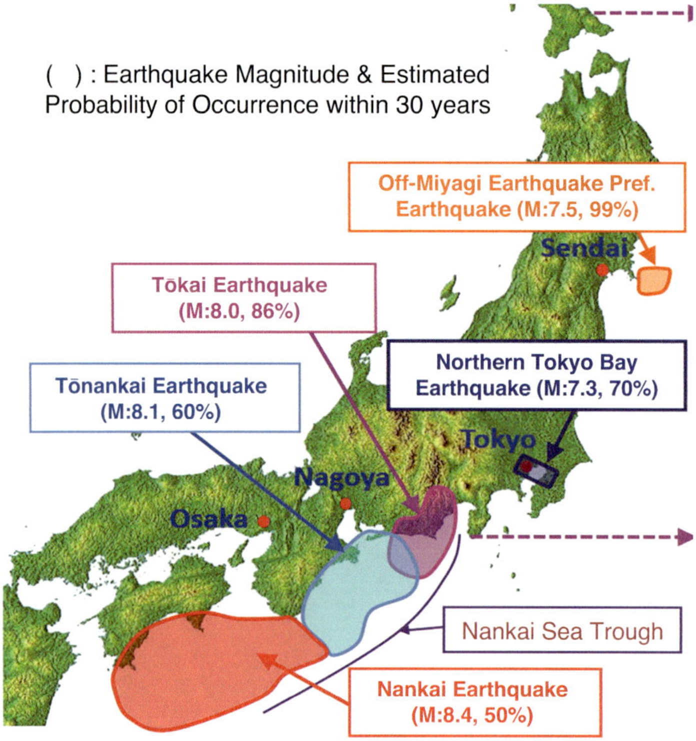

FIGURE 1.3.2
Source area of earthquake along the Nankai Sea Trough predicted before the 2011 East Japan earthquake.

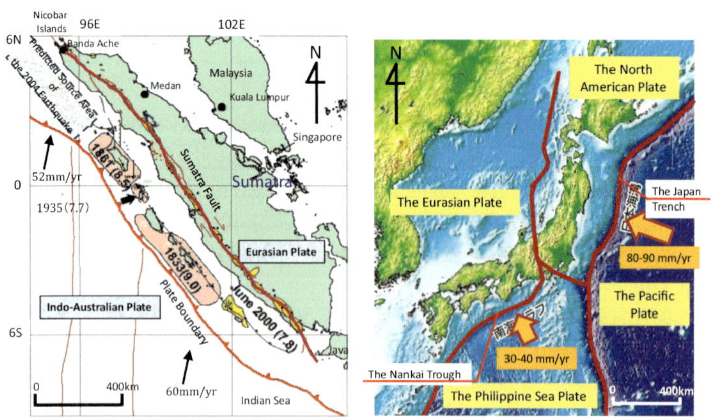

(1) The Plate Structure of the West Coast of Sumatra Island

(2) The Plate Structure of the Japanese Archipelago

FIGURE 1.3.3
Plate structures and earthquake generation mechanism of the west coast of Sumatra Island and the Japanese archipelago.

archipelago sits atop the North American Plate, with the Pacific Plate subducting beneath it. On the contrary, Sumatra Island lies on the Eurasian Plate, with the Indo-Australian Plate subducting beneath it. Rotating the plate structure off the coast of Sumatra by approximately 90° clockwise nearly aligns it with the plate structure of eastern Japan. It's puzzling why, despite this similarity, there was a belief that a mega earthquake of magnitude 9 or above would not occur in Japan.

The tsunami attacked the coastal area along the Pacific Ocean in Eastern-North of the Japanese main island, where the maximum run-up height reached about 30 m. The water height on shore was over 10 m as shown in Figure 1.3.4.

The number of totally and halfway collapsed houses and buildings was about 400,000, and over 700,000 houses and buildings were partially damaged. However, some buildings survived from the tsunami as shown in Figure 1.3.5. In the case of the 2004 Sumatra tsunami, some mosques at Banda Ache also survived. Those undamaged buildings showed that it was possible to construct structures that could resist against tsunami.

As a nation renowned for its leadership in natural disaster prevention in the world, Japan faced another significant earthquake calamity after the 1995 Kobe earthquake. As a researcher who has focused on earthquake disaster prevention for over half a century, the East Japan earthquake was another deeply regrettable disaster.

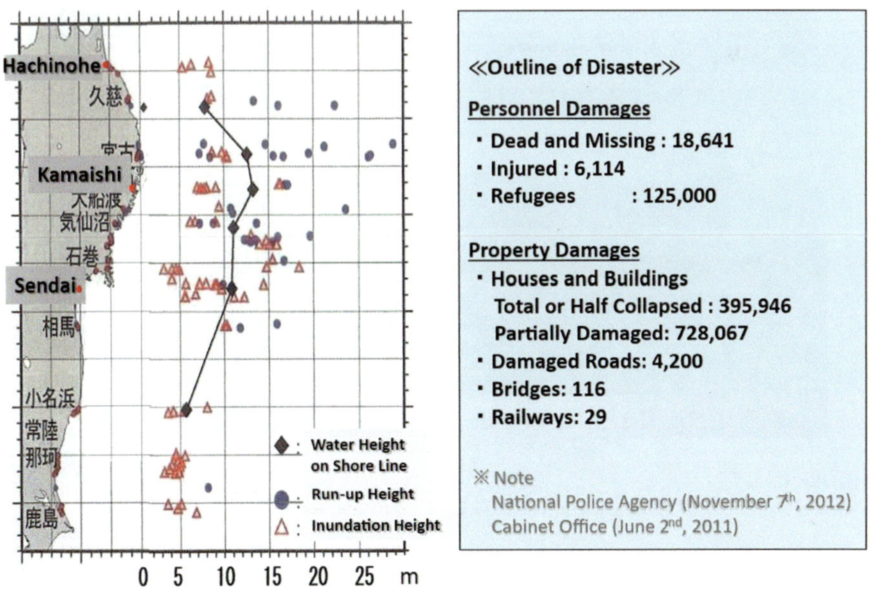

FIGURE 1.3.4
Tsunami height along the Pacific coast of the Tohoku region.

FIGURE 1.3.5
Buildings that survived from the tsunamis (the 2011 East Japan earthquake and tsunami and the 2004 Sumatra tsunami).

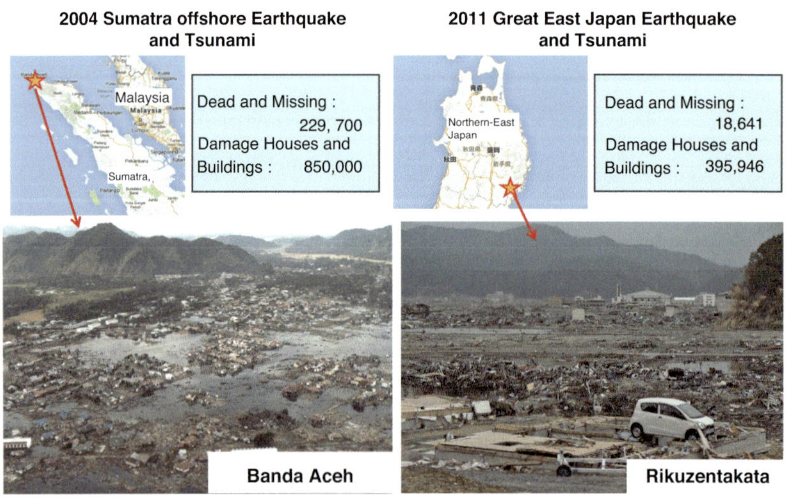

FIGURE 1.3.6
Disasters by the 2004 Sumatra and the 2011 East Japan tsunamis.

About a month after the disaster, I visited the affected area along the Pacific coast of the northeast area of Japan. The desolate scene in Rikuzentakata, where most homes and buildings were destroyed and swept away by the tsunami, reminded me of Banda Aceh in Sumatra, which was hit by the Indian Ocean earthquake and tsunami in 2004, killing about one-third of its population or 70,000 people. The similarity of the scenes of disasters of Rikuzentakata and Banda Aceh, shown in Figure 1.3.6, shocked me again as an expert in earthquake engineering.

I wondered why we had not foreseen a disaster like the one in the Indian Ocean occurring in Japan and urged for urgent measures. I had the same thought at the time of the 1995 Kobe earthquake. Despite having a duty to contribute to reducing natural disasters through our research in the field of earthquake engineering, we could have done better.

1.3.2 Critical Accident at Fukushima Daiichi Nuclear Power Plant[6]

Approximately 40 minutes after the East Japan earthquake on March 11, 2011, the Fukushima Daiichi Nuclear Power Plant was hit by the tsunami, leading to an unprecedented nuclear accident involving reactor core meltdown, hydrogen explosions, and collapse of reactor buildings. The tsunami height for the original design was 6.1 m, but the tsunami reached over 15 m in front of the site.

Damage to electronic breakers and the collapse of transmission towers due to slope failures outside of the plant caused all six external power lines to fail, prompting the emergency diesel generators for each reactor to start. However, the tsunami flooded the seawater pumps for cooling, emergency diesel generators, and distribution panels for Units 1–5. Except for Unit 6, which avoided flooding, all emergency diesel generators stopped, resulting in the loss of all AC power sources. Attempts were made to cool the reactors of Units 1–3 using batteries, but the batteries depleted, halting core cooling functions. Subsequently, efforts shifted to injecting freshwater and seawater through the fire extinguishing line using fire pumps, but difficulties arose due to increasing pressure in the reactors, making it impossible to inject water into the reactor pressure vessels. The nuclear fuel in the reactor of each unit became exposed and melted down.

A large amount of hydrogen was generated in the reactor containment vessel due to the chemical reaction between zirconium, which is part of the fuel rod cladding exposed in the atmosphere, and steam. Furthermore, the damage to the seawater system equipment led to the loss of the cooling system (final heat sink lost condition), causing damage to the reactor containment vessel. In Units 1 and 3, explosions occurred due to the ignited leaked hydrogen, destroying the uppermost floor of the reactor buildings.

The explosions of the reactor buildings dispersed substantial amounts of radioactive materials across a wide area, leading to radioactive contamination and forcing many residents to evacuate. More than one decade has passed since the accident, and although evacuation orders have been lifted in some areas, many areas remain under evacuation. Even if evacuation orders are lifted, many residents have built new lives in their places of evacuation over more than a decade, making it difficult for the population of the affected areas to return to pre-disaster levels.

Decommissioning work at the Fukushima Daiichi Nuclear Power Plant has continued for over a decade. I had the opportunity to visit the plant in 2020 to observe the progress of the decommissioning efforts. Around 3,000 workers were working day and night toward decommissioning tirelessly. The work progressed to the situation where it was possible to approach the reactor buildings without wearing radiation protection suits, but the road to decommissioning the nuclear reactor remains long and challenging.

There are two major challenges in decommissioning. The first challenge is how to remove the melted nuclear fuel that has accumulated at the bottom of the reactors. Photographing the condition of the melted fuel with cameras took ten years, and it may take several decades or more to remove and dispose of it externally. This issue will not be resolved within our generation but will be passed on to our children, grandchildren, and great-grandchildren, leaving a significant negative legacy for future generations.

Another challenge is the treatment of contaminated water and soil. Although the inflow of groundwater into the vicinity of the reactor buildings has been stopped by a frozen wall, contaminated water is generated every time it rains. This water has been stored in storage tanks for over ten years since the accident, but in 2023, the release of treated water into the ocean began. The treated water is released into the sea through a tunnel constructed from land toward sea. Efforts to gain the understanding of residents, especially those involved in the fishing industry, regarding the release of this treated water into the ocean are ongoing. Additionally, it is necessary to continue providing explanations to neighboring countries based on scientific data.

Not only in Fukushima Prefecture but also across a wide area of the northeast region of Japan, the soil has been contaminated with radioactivity. To rehabilitate the living environments of people and to revive agriculture and forestry, radioactive materials contained in the soil must be removed. Currently, these contaminated soils are still accumulating in the affected areas, and full decontamination has not yet started (Figure 1.3.7).

1.3.3 Development of Tsunami-Resistant Measures

The East Japan earthquake significantly eroded public trust in the science and technology for disaster prevention, with failures in earthquake and tsunami prediction, inadequate tsunami countermeasures represented by the severe accident at the nuclear power plant, liquefaction-induced housing damage, and fires and explosions in coastal industrial complexes all contributing to increased skepticism toward science and technology. One of the most important requirements for creating a "safe society" is people's trust in science and technology, which was severely undermined by the earthquake and the tsunami.

With the anticipated Nankai Trough mega-earthquakes, strong tremors are expected from the Tokai region to Kyushu, a 10,000 km distance, along with tsunamis exceeding 20 m in height. The question of how to prepare for such

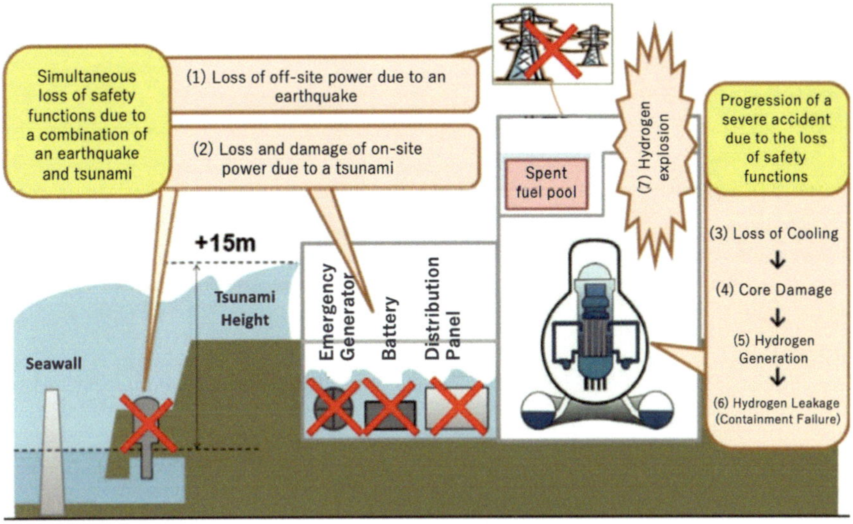

FIGURE 1.3.7
The accident at the Fukushima Daiichi Nuclear Power Plant caused by the tsunami[6]. (Nuclear Regulation Authority of Japan website.)

a massive tsunami is pressing. To restore public trust in disaster prevention science and technology, it is imperative to develop "tsunami engineering" in addition to traditional "earthquake engineering" and to advance tsunami countermeasures based on the outcomes from this new field.

The concept of "tsunami engineering" emerged from damage investigations of the 2011 East Japan earthquake and the 2004 Indian Ocean tsunami off the coast of Sumatra. During the East Japan earthquake, numerous houses, buildings, and bridges were destroyed by the tsunami, but amidst the devastation, some structures withstood the tsunami forces. Similar scenes were observed after the Indian Ocean tsunami, where Islamic mosques and concrete bridges near the coastline of Banda Aceh survived tsunamis over 10 m high, demonstrating that it is possible to construct structures that can structurally withstand tsunamis, as shown in Figure 1.3.5.

Furthermore, disaster education and tsunami evacuation drills conducted prior to the East Japan earthquake saved the lives of many school children. Measures from both hardware and software perspectives are required to mitigate future tsunami disasters (Figure 1.3.8).

After the East Japan earthquake, tsunami barrier walls have been constructed at all nuclear power plants in Japan. The critical point for the design is the estimation of the future tsunami height at each plant location. Furthermore, for increasing the safety of nuclear power plants, the tsunami measures such as watertight doors for reactor and generator buildings and the emergency equipment have been moved to a higher location of the plant (Figure 1.3.9).

FIGURE 1.3.8
Tsunami barrier walls of the Hamaoka Nuclear Power Plant (constructed after the 2011 East Japan earthquake).

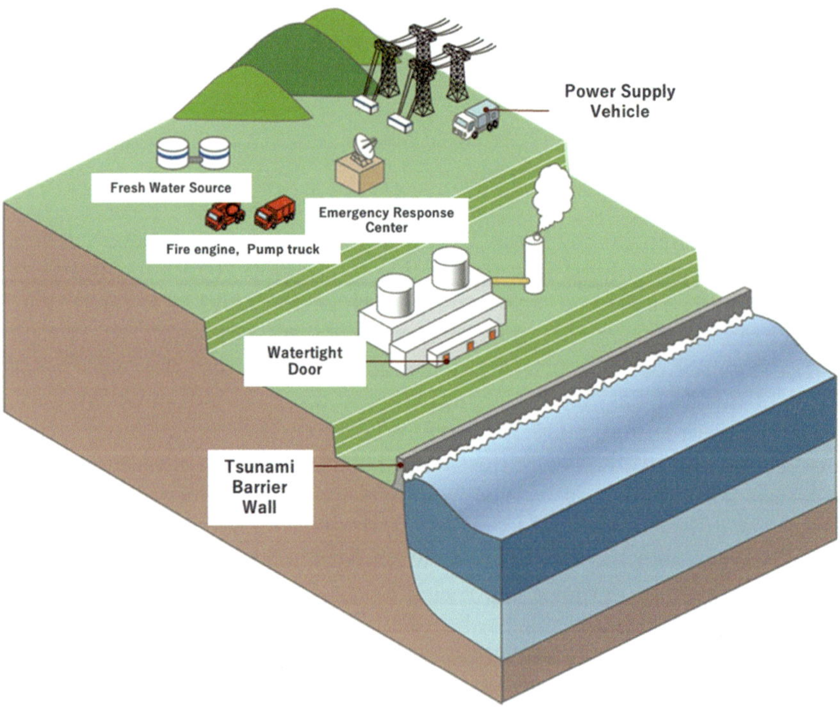

FIGURE 1.3.9
Tsunami measures at nuclear power plants. (FEPC's web magazine *Analog*, Vol. 3, June 2013.)

The East Japan earthquake, including the severe accident of nuclear power plant, has left a significant negative legacy for future generations. It is imperative not to repeat such disasters in the face of the imminent mega-earthquakes and tsunamis along the Nankai Sea Trough. To this end, the author proposed the following six initiatives to develop "tsunami engineering" and advance tsunami countermeasures:

i. *Investigate the global history of tsunamis from a geological perspective*: One of the reasons for the failure to predict the tsunami caused by the East Japan earthquake was the focus on historical documents spanning only a few hundred years. To understand tsunamis recurring over thousands of years, it is essential to conduct worldwide investigations of tsunami deposits from a geological perspective and share this information worldwide as a basis for tsunami prediction.

ii. *Construct social infrastructure and buildings resistant to tsunamis*: Despite being few, some buildings and civil engineering structures survived the tsunami during the East Japan earthquake. By scientifically clarifying the external forces generated by the tsunami, it is possible to construct buildings, bridges, and disaster prevention facilities that can withstand tsunamis.

iii. *Promote tsunami-resistant urban development*: Based on numerical analyses of tsunami behavior on land, tsunami-resistant urban development can be promoted through the selection of residential areas, street design, hillside land development, and the establishment of tsunami evacuation routes and shelters.

iv. *Improve tsunami resilience and rapid recovery of lifeline systems and water–front industrial parks*: The East Japan earthquake caused extensive damage to many lifeline systems such as sewage and industrial facilities like energy production plants, due to tsunami forces, collisions with floating debris, and flooding. Measures should be taken to maintain functionality and ensure rapid recovery from such impacts.

v. *Establish a system for rapid assessment and communication of disaster realities over wide areas*: To prepare for future widespread natural disasters, it is crucial to develop a system for collecting and transmitting damage information for lifesaving, emergency response and urgent recovery efforts.

vi. *Promote disaster education and training*: In Kamaishi City, Iwate Prefecture, and Kesennuma City, Miyagi Prefecture, the death and missing rates of school children due to tsunamis were one-tenth of the overall population rates, highlighting the effectiveness of disaster training and evacuation drills conducted continuously at primary schools before the tsunami, as shown in Table 1.3.1. The effects of disaster education and training should be examined, including the failure examples, to consider the best approach to disaster education and training.

The development of "tsunami engineering" and the advancement of tsunami countermeasures require interdisciplinary collaboration not only in natural

TABLE 1.3.1

Effects of Disaster Prevention Education in Kamaishi City and Kesennuma City

	Kamaishi City	Kesennuma City
Death and missing rate of the overall population	$\frac{1,091}{39,508} = 2.78\%$	$\frac{1,407}{74,247} = 1.89\%$
Death and missing rate of pupils and children	$\frac{5}{3,244} = 0.15\%$	$\frac{54}{6,054} = 0.19\%$

science and engineering but also in the humanities, social sciences, economics, law, information sciences, and emergency medical care. Therefore, academic societies related to tsunami disasters are urged to strengthen their collaboration further.

1.4 Liquefaction-Induced Large Ground Displacements[7,8]

1.4.1 Beginning of Research on Liquefaction-Induced Large Ground Displacements

On May 26, 1983, a moment magnitude 7.7 earthquake, the Japan Sea earthquake, occurred about 100 km off the coast of Akita Prefecture on the main island of Japan, causing damage primarily in the cities of Akita and Noshiro. The damage was particularly severe to the buried lifelines in the ground due to soil liquefaction.

In 1983, I retired from a construction company after 15 years of service to join the Faculty of Marine Science at Tokai University. Moving to the university, I encountered a significant difficulty: lack of research funds. Faculty members, especially in engineering schools, need funds for research activities and maintaining their laboratories. They should shoulder travel expenses and payment for the part-time work of students assisting in experiments and surveys.

Therefore, I contacted a friend working at Tokyo Gas company, with whom I had maintained a connection since my working at the construction company. I suggested, "The lifelines have suffered considerable damage due to the Japan Sea earthquake. I want to conduct a survey focusing on the damage to gas pipes. Could your company provide travel expenses?" There was no response for a while, but finally, in December of that year, the gas company committed a research theme titled "Study on the Damage to Gas Pipes Caused by 1983 Japan Sea Earthquake." I quickly started preparing for the survey with my students, but I was informed by the gas company that two engineers would accompany me. Being a new university faculty member, I suspected they were sent to ensure the survey was conducted successfully.

Finally, in late December, about seven months after the occurrence of the earthquake, we were able to depart for Akita. We rented a car from Akita station and headed for Hachirogata, where severe liquefaction had been reported. However, by the time we arrived, the snow had started to fall, covering the ground in white, making it impossible to identify traces of the soil liquefaction like sand boils and ground fissures. The wind grew stronger, and the cold became more severe, making it impossible to conduct the survey.

As evening approached, we decided to stay overnight at a nearby Japanese inn with a hot spring. Since the two employees from the gas company were both fond of Japanese sake, we ended up enjoying Akita's local sake that night, completely forgetting about the commission (Figures 1.4.1 and 1.4.2).

The next morning, we visited the Gas and Water Works Department in Noshiro City to investigate the damage. Two photos were shown to us (Figure 1.4.3), where a gas pipe at a 45° bend was completely ruptured, and the rupture ends were separated by 70 cm. Another photo showed a pipe that had once broken and was biting into each other. When asked by the staff in Noshiro City, "Why did this happen?". I was at a loss for an answer. I had assumed that the amplitude of ground displacements due to intense ground motions would be at a maximum of 10 cm, so the idea of a 70 cm separation was incomprehensible to me. These two photos became the beginning of the research on liquefaction-induced ground displacements.

FIGURE 1.4.1
Survey on liquefaction-induced ground displacement after the 1983 Japan Sea earthquake.

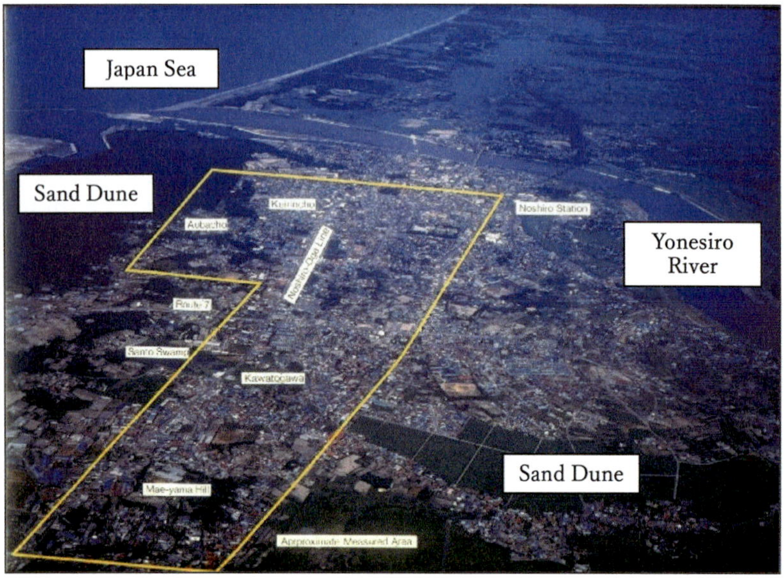

FIGURE 1.4.2
Aerial view of Noshiro City.

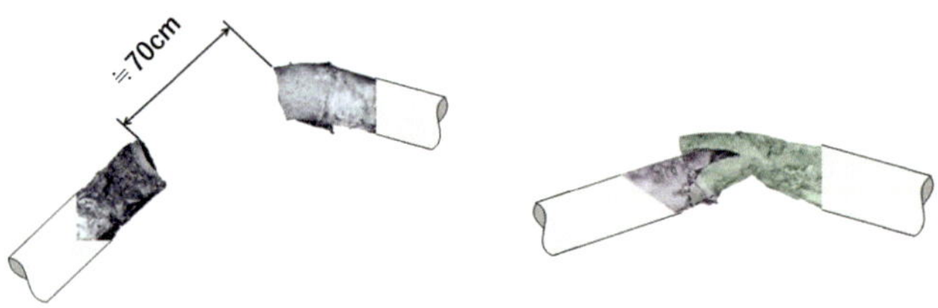

FIGURE 1.4.3
Rupture of gas pipes caused by the 1983 Japan Sea earthquake.

1.4.2 Measurements of Ground Surface Displacements[9–12]

Photographs of Noshiro City's urban area taken after the earthquake revealed roads that appeared to have been straight before the earthquake but were significantly curved horizontally afterward. Our ground surface observation also found cases where the boundary stones of residential properties had moved to almost the center of the garden after the earthquake, suggesting horizontal ground surface movement, possibly by several meters.

How to measure the ground displacements caused by the earthquake? In areas where ground displacement seemed significant, triangulation and plane table surveying were carried out. The idea was to redraw the post-earthquake city map through the surveying and compare the survey results with the pre-earthquake map to derive the ground displacements quantitatively. However, this method failed. The accuracy of the post-earthquake city map produced by the survey was low, and the measured results did not align with reported ground fissures and other changes of ground surface caused by the earthquake.

At this point, an employee from an aerial survey company suggested us to measure the ground displacements by aerial photogrammetry. This method involves using aerial photographs of the same area taken before and after the earthquake to measure the difference in three-dimensional coordinates of targets fixed on the ground surface. Targets for surveying included manhole covers, corners of sewer open channels, and foots of utility poles. In cases where appropriate targets were not available on the ground surface, corners of roofs of houses that were confirmed not to be tilted and moved horizontally due to the earthquake were selected as targets.

The aerial photogrammetry revealed up to 5 m of horizontal surface displacement on the Maeyama area in the southern part of Noshiro City. Maeyama, originally a sand dune area located away from the current coastline, is a gently sloping area with a gradient of 1%–2%. The survey revealed horizontal movements of the ground surface down the slope. When the survey company submitted its report, I initially doubted the results, finding it hard to believe that the ground could experience such extensive horizontal displacement due to an earthquake. However, looking at the survey results near where the gas pipe had ruptured and the two ruptured ends were separated by 70 cm, it became clear that the ground displacement occurred around the about 45° bent section, pushing outward across the entire pipeline with a magnitude of 2–3 m. These ground displacements caused large tensile force and bending moment at the bent part of the pipe, leading to the rupture and subsequent separation of the fracture ends.

1.4.3 Reaction of the Technical Committee of the Japan Society of Civil Engineers

The Earthquake Engineering Committee of the JSCE plays a role as a venue for exchanging information on earthquake damage and the seismic resistance of civil structures among engineers and researchers in the earthquake engineering field. At the committee's regular meeting, the results of ground displacement measurements in Noshiro City were reported, which are the survey results around Maeyama in Noshiro City as shown Figure 1.4.4. However, most members of the committee, particularly the researchers in the geotechnical field, did not believe the result of the measurement and advised that "the survey results are unbelievable. It must be a mistake in the measurement, so you should reexamine the result." Many other committee members appeared skeptical about the survey results. At that time, I was working at the Marine Science Department of Tokai University and was also affiliated with the Association for the Development of Earthquake Prediction Research,

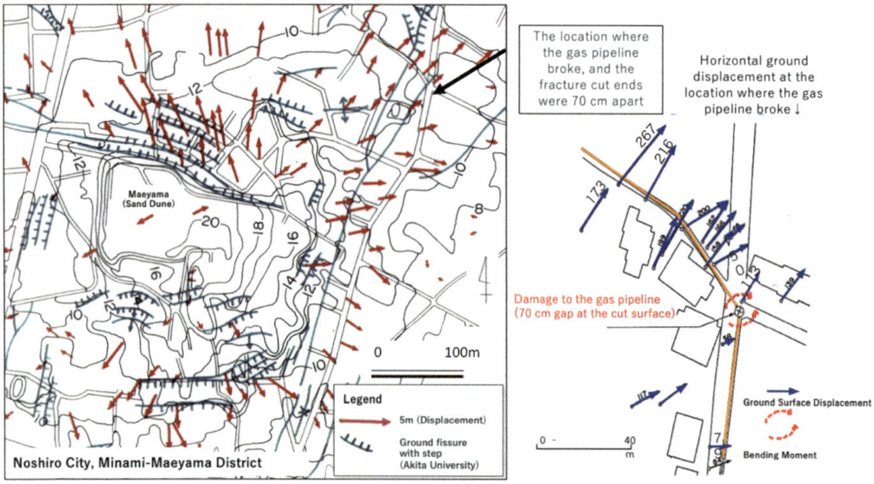

FIGURE 1.4.4
Horizontal displacements on the ground surface measured by aerial photogrammetry.

where Professor Tsuneji Rikitake, a prominent geophysicist, was at the promotion association. I reported the measurement results of the ground displacement. After listening to the author's explanation in silence, the professor first said, "The earth does move, doesn't it?" For a geophysicist, there was no question about the ground moving several meters. When research activities are conducted only within a limited field, one tends to no longer question the "common sense" of that specialized field. The fact that several meters of displacement of liquefied ground was not accepted by many researchers in the civil engineering field is an example of this.

1.4.4 Ground Displacements by the 1964 Niigata Earthquake[13]

Before the 1983 Japan Sea earthquake, similar events had occurred in past earthquakes where liquefaction took place, such as the 1964 Niigata earthquake. This earthquake led to a comprehensive investigation into liquefaction–induced ground displacements. In the Niigata earthquake, soil liquefaction occurred almost throughout the entire city along the Shinano and Agano rivers, causing damage such as the collapse of Showa Bridge constructed about a month before the earthquake, damage to a huge length of lifeline buried pipelines, and tilting and sinking of buildings and houses. It was the first earthquake where the phenomenon of soil liquefaction and its caused damage were recognized from an engineering perspective.

Surveys by aerial photographs taken before and after the earthquake showed that the ground along the Shinano and Agano rivers had moved horizontally by several meters, as shown in Figure 1.4.5. On the right bank of the Shinano River, a maximum horizontal displacement reached 10 m.

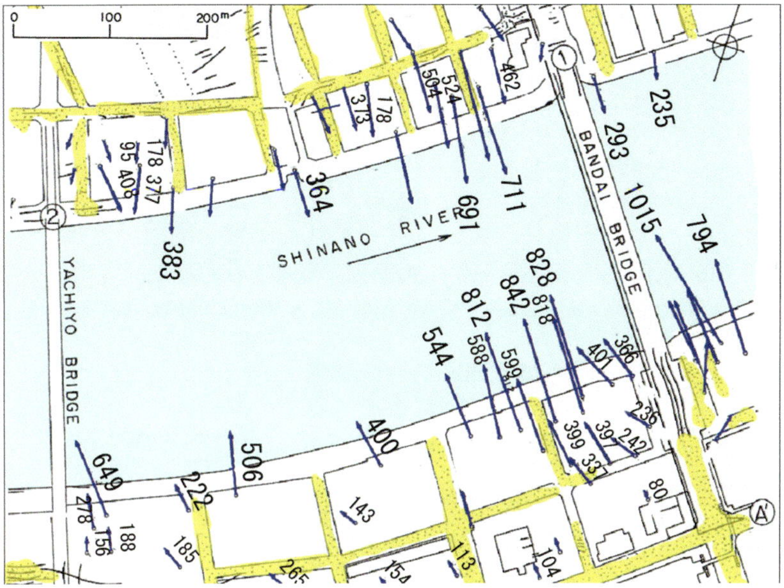

FIGURE 1.4.5
Liquefaction-induced ground displacement along the Shinano River (cm).

These survey results were once again reported at the JSCE committee, but the reaction of the committee members was the same as before. The comment from the committee was that "It is impossible for the ground to move several meters, let alone more than 10 meters, due to liquefaction. It must be a mistake in the survey, again."

Clear evidence was needed to convince the researchers and engineers that the ground had largely moved horizontally. Was there anything left even after more than 20 years since the earthquake occurred? The following summer in 1985, after conducting aerial photographic surveys, an engineer from Tokyo Gas and I went to Niigata City to find evidence of horizontal ground movements. We walked around Niigata City all day but could not find any clear traces of ground displacements. Both of us tiredly returned toward Niigata station on the left bank of the Shinano River. At that time, we noticed an abnormality in the road along the left bank of the Shinano River. Near the bridge abutment, the embankment and road were unnaturally curved. Near the bridge abutment, the embankment was caved in toward the abutment side. This would have been obvious if we had carefully examined the aerial photographs taken before and after the earthquake. But we overlooked this change in the embankment. After returning from Niigata, we looked at the aerial photographs again. Clearly, the embankment and road were bent after the earthquake as shown in Figure 1.4.6. The brick walls behind the embankment along the river were also unnaturally bent. Near the bridge abutment, ground displacement was reduced due to the rigidity of the foundation of

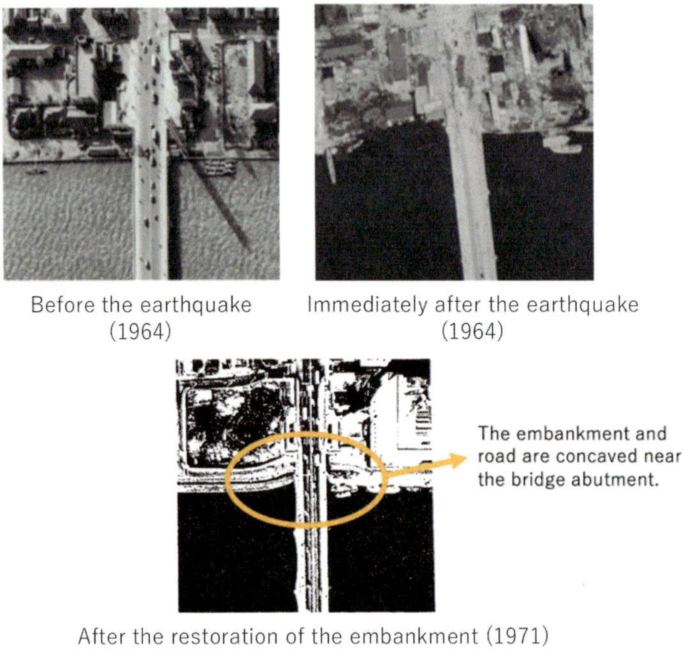

FIGURE 1.4.6
The transition of the left bank line of the Bandai Bridge

the abutment, but further away from it, the ground had moved significantly toward the center of the river (Figure 1.4.7).

While measuring ground displacement in Niigata City, Dr. Kawamura, a former colleague at a construction company, visited my office of the Association of Development of Earthquake Prediction (ADEP) in Tokyo and said, "While investigating the damage to buildings caused by the Niigata earthquake, I found something strange." When excavating the foundation ground to rebuild the building that had tilted and sunk due to the Niigata earthquake, it was found that all the concrete piles with a diameter of 30 cm had been broken. The breakage of the concrete piles occurred at two elevations. Although the building had tilted and sunk due to the earthquake, it had been used for about 20 years by replacing the inclined floor to make it horizontal. This building was three-stories NHK broadcasting station, and Mr. Yanagawa Yoshiro, an acquaintance of the author who was a commentator at NHK said, "Whenever I went on a business trip to the Niigata broadcasting station, I always felt wobbly, especially on the landings of the stairs."

Twenty-five years after the Niigata earthquake, aerial photographic surveys revealed that the liquefied ground had been horizontally displaced by as much as 10 m, but right after the earthquake, many citizens of Niigata City had the impression that the ground had moved significantly.

(a) It was discovered about 20 years after the Niigata earthquake that all the foundation concrete piles of a building had been broken due to liquefaction-induced ground displacement.
(b) The piles were broken at two elevations.

FIGURE 1.4.7
Concrete piles destroyed by liquefaction-induced ground displacement

About two months after the earthquake, at a "Citizens' Forum on the Niigata Earthquake," three citizens made the following statements:

Mr. A: My house is about 70 m from the left bank of the Shinano River. While the house itself was not damaged, when I measured the dimensions of my lot, it stretched by about 2.7 m. This is because the ground, of wide areas and houses, has moved toward the Shinano River.

Mr. B: Today, as I crossed the Yachiyo Bridge on the Shinano River to come to this venue, I indeed felt as though the land along the river embankment had extended toward the river, making the land appear wider.

Mr. C: Before the earthquake, the hedge between my house and the neighbor's house at the back of my property was narrow, and I could not pass through, but after the earthquake it was expanded to allow passage. I am contemplating what could have caused this.

Researchers first noticed the significant ground displacement due to soil liquefication at the time of the 1983 Japan Sea earthquake, about 20 years after the Niigata earthquake. If researchers had paid attention to such statements from the citizens, studies on liquefaction-induced ground displacement might have advanced. The 1995 Kobe earthquake also caused large ground displacements due to liquefaction, leading to bridge girder falls and damage to lifeline pipelines. Had countermeasures against the ground displacement been implemented, these damages could potentially have been prevented (Figures 1.4.8 and 1.4.9).

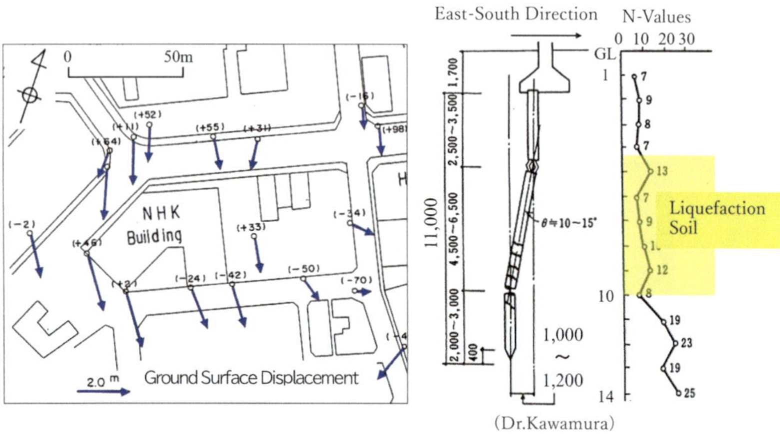

FIGURE 1.4.8
Deformation of damaged piles and ground surface displacements around the building.

FIGURE 1.4.9
The collapse of Showa Bridge by liquefaction-induced ground displacements.

Among the damages to infrastructures caused by the Niigata earthquake, the collapse of the Showa Bridge is noteworthy. This bridge had just been constructed about one month before the earthquake for the National Sports Festival held in Niigata City. The bridge girders moved from the piers and fell into the river. There are two theories regarding the cause of the falling of the girders. One theory suggests that the piers vibrated significantly due to the strong earthquake motion, causing the distance between two adjacent piers to widen and the simply supported girders to fall. The other theory posits that the ground of the river bed liquefied, leading to horizontal displacements, which pushed and deformed the foundation of the piers. If the first theory is correct, the bridge would have collapsed while the earthquake shaking was ongoing, but there exists a taxi driver's testimony that contradicts this explanation. The driver said: "I was driving my car and felt the earthquake ground motion when I was almost in the middle of the bridge. I stopped my car and waited for the shaking to subside, which took a few minutes. After the shaking stopped, I got out of the car and ran toward the left bank. When I looked back, the bridge girders I had just passed were falling into the river one after another." If this testimony is dependable, it suggests that the falling of the girders began after the shaking had stopped. A ground motion by the Niigata earthquake was recorded at the Niigata Meteorological Observatory, which shows that the main shaking lasted only about 30–40 seconds.

Another theory for the falling of the bridge girders, which I advocate, is the theory of liquefaction-induced ground displacement. Liquefaction, as the term implies, refers to the ground behaving like a liquid. Liquefaction occurs when the pores between sand particles are saturated with water, causing the pore water pressure to increase, the contact between sand particles to break, and the particles to separate in the water, a state referred to as liquefaction. Liquefaction and horizontal displacement continue until the pore water pressure decreases and the contact of the sand particles returns to its original state. It is believed that in the case of the Niigata earthquake, the ground remained in a liquefied state for possibly ten minutes. Based on the taxi driver's testimony, the falling of the girders required at least a few minutes, leading to the conclusion that liquefaction-induced ground displacement, rather than the earthquake shaking, was the cause of the Showa Bridge's collapse.

In Niigata City, it was found that many roads had been bent due to liquefaction-induced ground displacements. Figure 1.4.10 was one of the

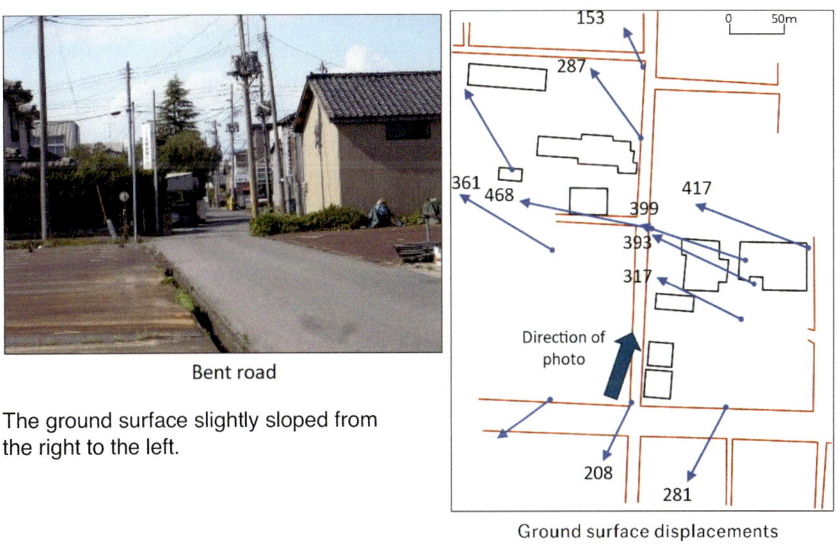

FIGURE 1.4.10
Road bent by liquefaction-induced ground displacement.

examples of such roads. Before the earthquake, the road had been straight. The ground surface was slightly inclined from right to left, and the ground moved downwards.

1.4.5 Japan-US Joint Research[8]

Around the same time, when research on liquefaction-induced ground displacement in Japan began following the 1983 Japan Sea earthquake, similar studies were underway at Cornell University in the United States, led by Professor Thomas O'Rourke. Their studies included research on the 1906 San Francisco earthquake, where liquefaction in the downtown area around Mission Creek caused up to 3 m of horizontal ground displacement.

For the case studies of the 1906 San Francisco earthquake, aerial photos did not exist, so the photographs taken on the ground surface after the event were used for the study. The streets of San Francisco, arranged rectangularly, provided a unique opportunity to quantify ground displacements. It was noticed that streets photographed after the earthquake were no longer straight but bent, and the horizontal displacement of the ground was measured as the differences between curves and the original straight lines, referred to as offset displacements. After the 1964 Niigata earthquake, residents testified that many roads, which had been straight before the earthquake, were bent as shown in Figure 1.4.10, and the view ahead along the road was worsened.

The U.S. research findings were first presented at the 1986 U.S. Earthquake Engineering Conference in Charleston, South Carolina, which I attended. Professor O'Rourke already recognized the measurements of liquefaction-induced ground displacements during the 1983 Japan Sea and 1964 Niigata earthquakes published in a report by the Association for the Development of Earthquake Prediction. Similar researches were being conducted concurrently in Japan and the US.

On the last day of the conference, a closing party hosted by the Earthquake Engineering Research Institute, U.S., took place on a cruise ship touring Charleston Harbor. While enjoying an exhibition of the Charleston dance and sipping bourbon whiskey, Professor O'Rourke and I discussed Japan–U.S. collaboration for the study on liquefaction-induced ground displacements and their caused damage.

We agreed to hold "Japan–US Workshops" to report on research findings regularly and discuss collaborative research directions. The first workshop took place in Tokyo in 1988, followed by seven workshops in Buffalo, San Francisco, Honolulu, Snowbird (Utah), Tokyo, Seattle, and again in Tokyo (Figure 1.4.11).

Workshop topics included a wide range of themes related to soil liquefaction and its induced ground displacements, seismic design and retrofitting of lifeline facilities, liquefaction hazard assessment, and surface fault countermeasures. The papers presented at these workshops were published by the National Center for Earthquake Engineering Research, U.S. (NCEER).

FIGURE 1.4.11
Japan–US Joint Research Workshops (1st meeting in Tokyo).[8]

One outcome of the Japan–U.S. joint research was case studies compilation on liquefaction-induced ground displacements and their caused damage to lifeline facilities during past earthquakes, published by NCEER as "Case Studies of Liquefaction and Lifeline Performance During Past Earthquakes, Volume 1 (Japanese Case Studies), Volume 2 (United States Case Studies)." The Japanese case studies included the 1923 Kanto earthquake, the 1948 Fukui earthquake, the 1964 Niigata earthquake, the 1983 Japan Sea earthquake, and the 1990 Luzon, Philippines earthquake. The US case studies covered the 1906 San Francisco earthquake, the 1964 Alaska earthquake, the 1971 San Fernando earthquake, and several others (Figure 1.4.12).

For the 1971 San Fernando earthquake in California, pre- and post-earthquake aerial photos were used to measure ground displacements, and more than 3 m ground displacements were observed around Van Norman Lake in the San Fernando Valley.

Aerial photogrammetry for measuring ground displacements due to liquefaction was also applied to the 1999 Turkey Kocaeli earthquake and the 2010 New Zealand Darfield earthquake, in collaboration with researchers and stakeholders from Japan, the US, and New Zealand.

Model vibration experiments were conducted jointly by the University of California, San Diego and Waseda University to elucidate the mechanism of liquefaction-induced large ground displacements. A large experimental soil tank, 5 m long, was transported from Japan to the US, and experiments were

FIGURE 1.4.12
Shaking table test jointly conducted by UC San Diego and Waseda University. (Prof. Elgamal is at front center.)

conducted using the shake table at UC San Diego. Graduates and undergraduate students from Waseda University spent two months at UC San Diego, participating in experiments under the guidance of Professor Ahmed Elgamal.

In the case of the 1991 Kocaeli earthquake in Turkey, due to national security concerns, aerial photographs were prohibited from being taken outside the country.[14] Therefore, two Turkish Army officers (both reportedly colonels) brought the photos to Japan under an official permission of the army, and aerial photogrammetry was conducted under their supervision, marking a unique collaboration between Waseda University and the Turkish Army. This collaboration revealed significant liquefaction-induced displacements of up to 4 m in sandy ground on the east shore of Sapanca Lake, about 100 km east of Istanbul. A five-story hotel built on the coast of the lake moved and tilted due to the ground displacement (Figures 1.4.13 and 1.4.14).

1.4.6 The 2024 Noto Peninsula Earthquake and the Governmental Policy for National Land Resilience[15]

On the New Year Day of 2024, at 16:10, an earthquake with a magnitude of 7.6 occurred off the coast of the Noto Peninsula along the Japan Sea in Ishikawa Prefecture, resulting in approximately 500 victims (including related deaths

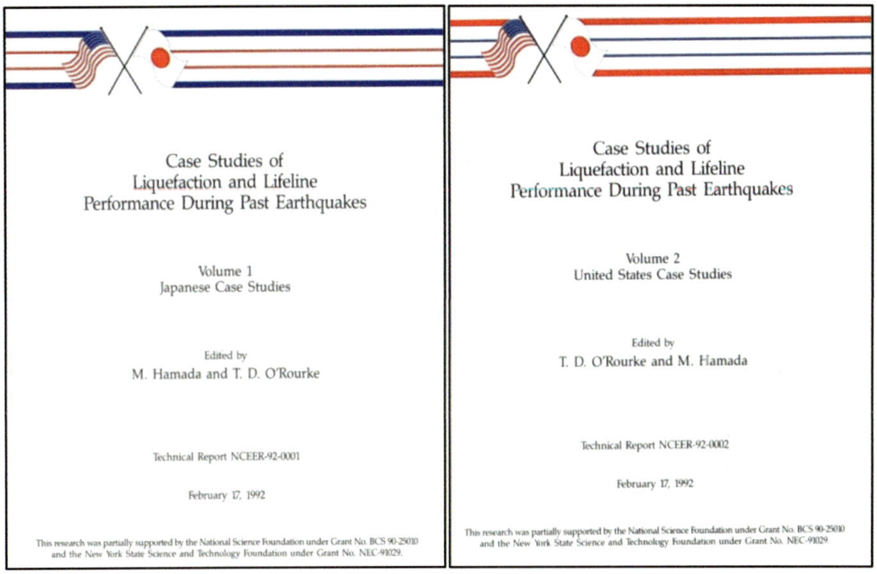
Case Studies of
Liquefaction and Lifeline
Performance During Past Earthquakes

Volume 1
Japanese Case Studies

Edited by
M. Hamada and T. D. O'Rourke

Technical Report NCEER-92-0001

February 17, 1992

This research was partially supported by the National Science Foundation under Grant No. BCS 90-25010 and the New York State Science and Technology Foundation under Grant No. NEC-91029.

Case Studies of
Liquefaction and Lifeline
Performance During Past Earthquakes

Volume 2
United States Case Studies

Edited by
T. D. O'Rourke and M. Hamada

Technical Report NCEER-92-0002

February 17, 1992

This research was partially supported by the National Science Foundation under Grant No. BCS 90-25010 and the New York State Science and Technology Foundation under Grant No. NEC-91029.

FIGURE 1.4.13
Report on case studies in Japan (six earthquakes) and the United States (five earthquakes). (National Center for Earthquake Engineering Research, NCEER-92-0001, Feb. 17, 1992.)

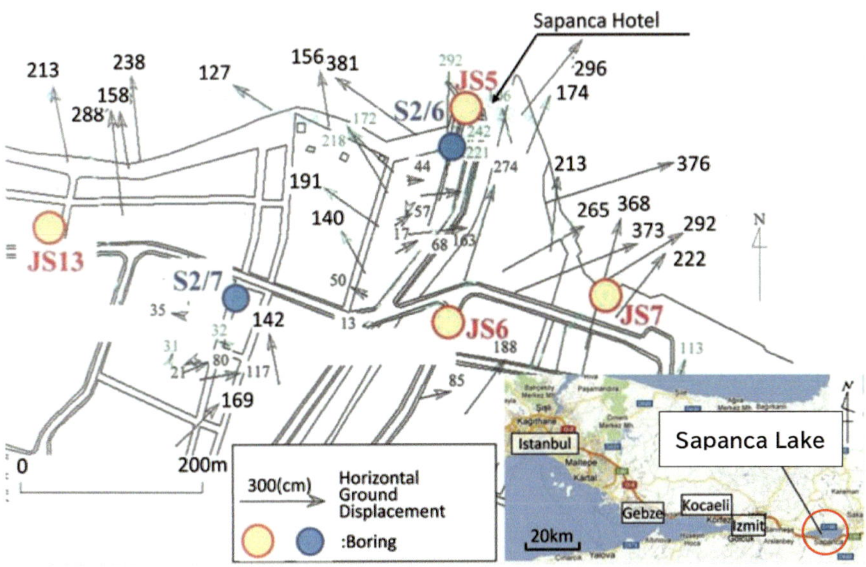

FIGURE 1.4.14
Turkey–Japan cooperative research on liquefaction and its caused large ground displacements in the 1999 Turkey Kocaeli earthquake.[14]

after the earthquake) and damage to 46,000 houses and buildings. The earthquake caused widespread damage, including a fire in the center of Wajima City, which exacerbated the disaster. The Japan Meteorological seismic intensity of VII, the highest rank, was recorded in Wajima City, and at Shika Town, about 70 km distant from the epicenter, a ground surface acceleration of about 1,800 cm/s^2 was recorded. This greatly exceeds the 810 cm/s^2 recorded in Kobe City during the 1995 Kobe earthquake.

The Noto Peninsula earthquake was caused by the ruptures of several undersea faults of 150 km length off the peninsula. In the case of this earthquake, no agency or seismologist could have predicted its occurrence. Predictions by the National Earthquake Research Committee indicated that the likelihood of an earthquake with seismic intensity is VI or higher occurring in the Noto region, but the estimated probability of the earthquake occurrence was the lowest nationwide. It has been considered impossible to predict the exact location and the time of earthquakes offshore. Regarding the Pacific coast, the risk of concurrent Tokai, Tonankai, and Nankai earthquakes has been warned as shown in Figure 1.3.2, and measures by the government and local authorities were being taken against these earthquakes. However, the actual mega earthquake occurred off the coast of the Tohoku region in 2011 as shown in Figure 1.3.1; any seismologist could not predict the earthquake. This fact clearly showed the people that the earthquake prediction was actually impossible at the present time (Figure 1.4.15).

Substantial national funds are invested annually in earthquake prediction research and observations. Many past earthquake disasters have highlighted the necessity of investing more in preventing the collapse of buildings and homes, which are directly related to the large loss of human lives, rather than in uncertain research and studies on earthquake prediction.

The most severely affected areas are located at the tip of the peninsula, and the damage to road function due to numerous slope sliding severely disrupted transportation and delayed emergency and recovery efforts. The damage to lifelines was also severe, with more than 6,400 homes among 13,000 houses in Wajima City without water supply for more than two months after the earthquake (as of March 1), causing difficulties for the affected people.

The 2024 Noto Peninsula earthquake also caused large ground displacements due to soil liquefaction. In the Uchinada area along the Japan Sea in Kanazawa City, a gently sloped area moved downward horizontally over 3 m.

Figure 1.4.16 shows horizontal displacements with ground failures like ground fissures, sand boils, subsidence, and rising up. The soil liquefaction and ground displacements caused severe damage to residential areas.

Approximately ten years before the Noto Peninsula earthquake, in December 2013, the Basic Act on National Land Resilience was enacted in the Japanese Diet. Its basic policies include the following: (i) protecting human life, (ii) maintaining the critical functions of the nation and societies, (iii) minimizing damage to the property of citizens and public facilities, and (iv) facilitating rapid recovery and reconstruction. By the fiscal year 2022, the

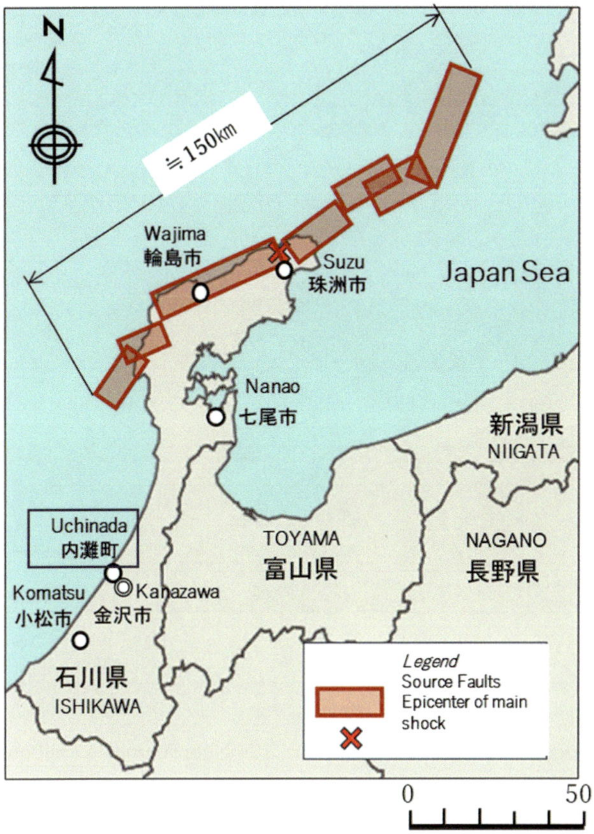

FIGURE 1.4.15
Source of the 2024 Noto Peninsula earthquake.

total government expenditure for national resilience had reached about 500 billion US dollars. These measures by the government, local authorities, and private businesses were believed to have steadily reduced the risk of natural disasters in Japan. However, the Noto Peninsula earthquake was a stark reminder that this belief was overly optimistic.

The primary cause of loss of lives by the earthquake was the collapse of houses and buildings, a recurring theme in past disasters such as the 1995 Kobe earthquake and the 2016 Kumamoto earthquake. Not only in Japan but globally, especially in Asia, the collapse of vulnerable houses and buildings is a significant cause of casualties by earthquakes. The earthquake near the Turkey-Syria border in January 2023 also resulted in considerable loss of human lives due to the collapse of buildings with low earthquake resistance.

Regarding the national policy on resilience mentioned above, the second principle after protecting human life is maintaining the vital functions of the

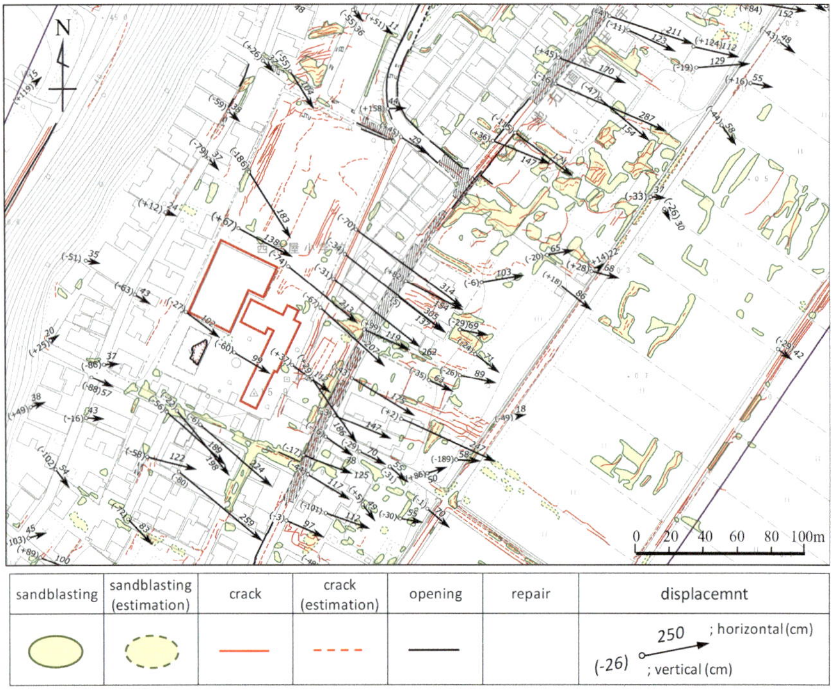

FIGURE 1.4.16
Ground deformation and failures caused by the 2024 Note Peninsula earthquake (Uchinada, Kanazawa City, Ishikawa).[15]

nation and the societies. However, measures have predominantly focused on urban areas along the Pacific coast, Tokyo, Nagoya, and Osaka areas. The extent of resilience efforts in depopulated areas, such as peninsulas and remote islands, is not enough. The Noto Peninsula earthquake provides an opportunity to review the reality of strategies for national land resilience and to reconsider priorities.

The lack of progress in seismic retrofitting, leaving many houses vulnerable, is primarily due to the financial burden on residents. Although local governments are developing subsidy programs for seismic retrofitting, there are still difficulties to prepare enough funds. There is also societal resistance to investing public funds in reinforcing private properties. Additionally, despite fearing a future earthquake, many people convince themselves that it won't affect them or their families, hindering progress in seismic retrofitting.

If vulnerable buildings and houses are not reinforced, tragedies like the Noto Peninsula earthquake will inevitably repeat. Technologically, there is a demand for the development of low cost retrofitting methods accessible to residents.

The effectiveness of seismic retrofitting on public buildings was demonstrated in the Noto Peninsula earthquake. Retrofitting had been carried out on school buildings and community centers, and although some damage occurred, these buildings could be used as primary evacuation shelters after the earthquake. Similarly, no significant damage was reported to civil engineering structures such as bridges and seawalls, indicating the success of resilience efforts.

Various agencies and researchers are expected to continue investigating the disaster and the causes of damage to structures. Despite experiencing unprecedented high accelerations of 1,800 cm/s^2, many structures and buildings either remained undamaged or maintained their functionality with only minor damage. Studies into the reasons why these structures could withstand such intense seismic activity are also required. The results of the studies are expected to serve as a basis for further strengthening and effectively advancing national land resilience.

1.4.7 Repeated Tragedy: Heavy Rain and Flood after the Noto Peninsula Earthquake

About eight months after the Noto Peninsula earthquake, August 21, 2024, heavy rains and floods ravaged the same area in Noto Peninsula where recovery and reconstruction had just begun from the earthquake disaster. In Wajima City, the maximum rainfall intensity per hour reached over 120 mm/hour, and the total volume of the rain for 3 hours recorded 220 mm, which largely exceeded the design level of the sewage system of the city. Furthermore, embankments of rivers collapsed, and the flood spread to huge residential areas. The rain flood killed 13 people in addition to 462 casualties by the earthquake.

After the earthquake, evacuation shelters were constructed in the affected areas and accommodated many victims, but a number of the evacuation shelters were flooded up to floor level. The victims had to move again to other places.

Due to the strong earthquake motions caused, numerous slope slides buried houses and severely damaged lifelines such as road, electricity, water, and sewer. In the landslide areas, many ground fissures remained after the earthquake without any restoration. A large amount of rainwater flowed into the ground fissures and caused more slope slides. The heavy rains and floods crushed the people's mind for recovery and reconstruction from the earthquake disaster. The risk of duplication of different kinds of disasters has been warned, but any concrete measure has not been realized (Figure 1.4.17).

1.5 Islands Sunk by Earthquakes

1.5.1 Uryu Island in Beppu Bay of Kyushu

In the spring of 1987, two local historians from Oita Prefecture of Kyushu visited the Faculty of Marine Sciences at Tokai University in Shimizu City. They were interested in investigating the truth behind the legend of Uryu Island,

FIGURE 1.4.17
Slope sliding caused by the earthquake (2024/1/1) and the heavy rainfall (2024/8/21): (a) slope sliding by the earthquake, (b) restoration of the road (2024/8), and (c) slope sliding by the rainfall (2024/9).

which was said to have existed in the Beppu Bay on Kyushu Island and sunk due to an earthquake. The story of Uryu Island, which supposedly sank into the bay along with about 1000 residents by a magnitude 7.3 earthquake on July 12, 1596, has been a subject of long-standing debate among local historians, primarily based on historical documents. Uryu Island was known to have been a prosperous port town, measuring approximately 3.5 km east to west and 2.5 km north to south.

The first known record of Uryu Island is found in the "Hofu Kibun," written about 100 years after the earthquake. This document describes the devastation caused by the earthquake and the subsequent tsunami in the Beppu Bay area, including the sinking of Uryu Island. However, there are significant doubts about the reliability of this account, mainly because it was written about one century after the event, based on hearsay, and no contemporary documents mention the sinking of the island.

A map of Uryu Island is included in the "Hoyo Kojiden," edited during the Edo period, more than 260 years after the earthquake (Figure 1.5.1). This map, detailed with capes, bays, and streets, suggests the island's extensive

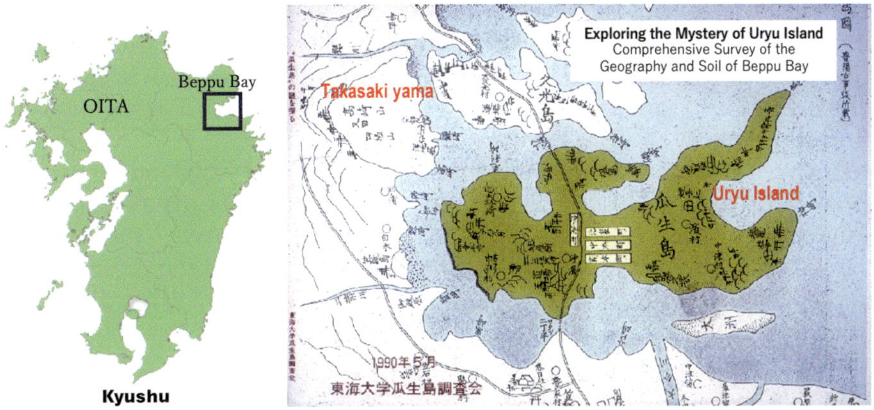

FIGURE 1.5.1
Old Map of Uryu Island "Hoyo Kojidan" (1857).

FIGURE 1.5.2
Research of Uryu Island by the Tokai University training ships.

presence in the Beppu Bay. However, depth measurements near Takasakiyama (Figure 1.5.1), where a part of the island was supposedly located, indicate depths of over 60 m, making it unlikely for an island to have sunk to such depths due to a magnitude 7.3 earthquake alone. However, several historical records mention Uryu Island or "Oki no Hama" (Beach off coast) before the 1596 earthquake (Figure 1.5.2).

These include accounts of Francis Xavier's visit to Bungo (present-day Oita City) 45 years before the earthquake, the records of the Chinese envoy Zheng Shungong, who was stranded in Bungo due to a typhoon in 1556, and a report by Luis Frois, a Jesuit missionary who described a catastrophic event at "Oki no Hama" in 1596.

To investigate the existence and possible sinking of Uryu Island, a research group from Tokai University conducted seabed surveys in the area believed to have been the location of the island. The sounding surveys did not find significant irregularities on the seabed that would indicate the presence of a sunken island (Figure 1.5.3). However, acoustic surveys revealed disturbed strata near the coast, suggesting large-scale ground failure possibly related to an earthquake and subsequent liquefaction (Figure 1.5.3).

Further investigations, including sediment core samples, showed orderly layers of sand deposits near the current mouth of the Oita River, which could have resulted from the disturbed sand layers resettling after being agitated by the earthquake (Figure 1.5.4).

Based on these findings, the research group speculated that Uryu Island might have been a sandbar formed by the Oita River and Ono River, which underwent significant subsidence and lateral movement due to liquefaction caused by the magnitude 7.3 earthquake. Some parts of the island might have remained above sea level after the earthquake but were gradually submerged due to high waves and tides over the following century (Figure 1.5.5). This theory provides a plausible explanation for the lack of contemporary

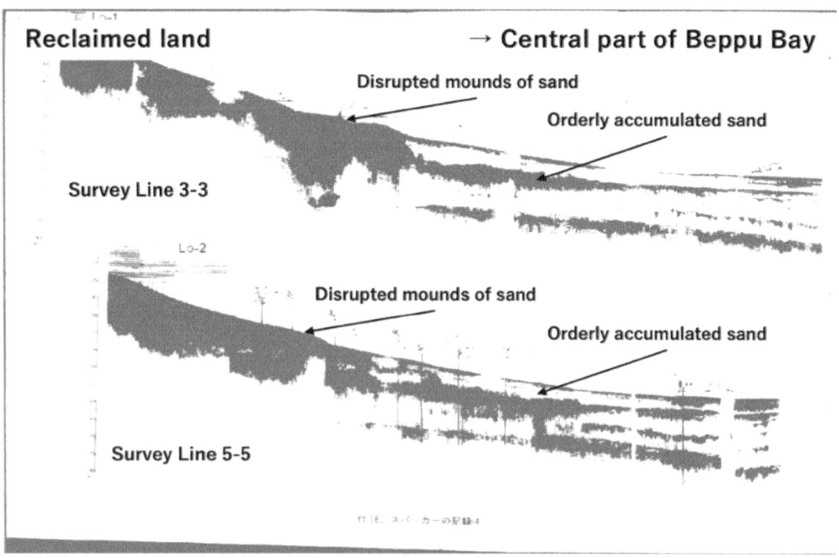

FIGURE 1.5.3
The record of the seabed acoustic survey.

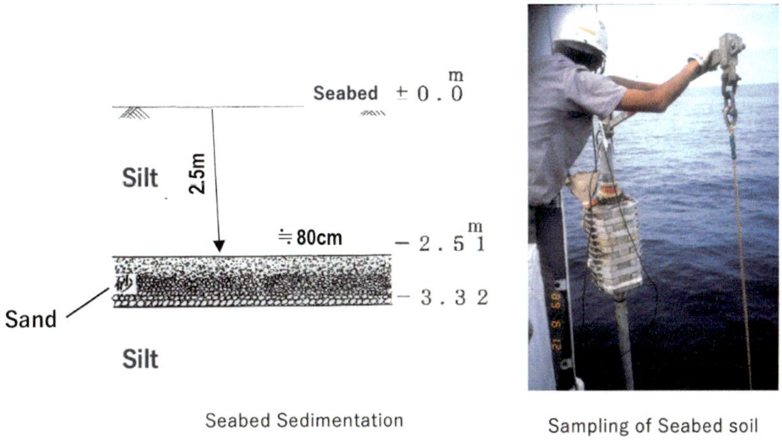

FIGURE 1.5.4
Survey of sedimentation of seabed by Piston Corer.

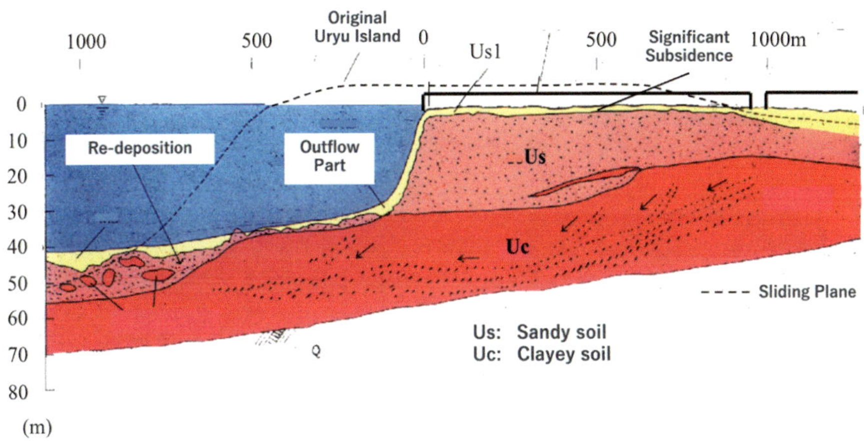

FIGURE 1.5.5
Estimation of the cause of Uryu Island's submergence (liquefaction and seabed slides).

records mentioning the sinking of Uryu Island for about 100 years after the earthquake.

This hypothesis was further supported by observations made after the 1990 Luzon Island earthquake in the Philippines, where similar large-scale ground movements and liquefaction were documented.

The seabed slopes gently away from the reclaimed land toward the center of the bay, and there are no traces of sunken islands. According to the acoustic survey, the geological layers are significantly disrupted near the reclaimed land. Offshore, the soil layers are deposited neatly.

At the seabed sedimentation conditions of about 100 m offshore from the mouth of the Oita River, about 2.5 m of silty soil is deposited from the seabed surface, and below that, sandy soil is deposited for about 80 cm. This layer of sand is composed of coarse to fine grains of sand, and it is presumed that the sand was once disrupted and then redeposited.

A large part of the island caused sliding due to liquefaction of sandy soil, but some parts remained above sea level with significant subsidence. It is assumed that the island continued to sink for 100 years due to high tides and big waves and finally disappeared below the sea surface.

1.5.2 Sinking of Sand Dunes in Philippines[16]

The moment magnitude 7.8 earthquake that struck the island of Luzon in the Philippines on July 16, 1990, caused significant damage to bridges, buildings, houses, and lifelines. The liquefaction damage along the coast of Lingayen Gulf in the northern part of Luzon was particularly severe.

Narvacan, a village with a population of about 1,000, experienced significant subsidence due to liquefaction, with parts of it submerged below sea level. The village was located on a sandbar developed along the western coast of Lingayen Gulf, and it was reported that an area approximately 1 km long and 500 m wide was submerged below sea level (Figure 1.5.6).

FIGURE 1.5.6
Lingayen Gulf of the Philippines (1990 Ruzon Island earthquake).

The village chief testified about the situation at the time, saying,

> The earthquake was so strong that it was impossible to stand, and it lasted for about a minute. Immediately after the earthquake, hot groundwater was spouting everywhere, and water columns rose in three places on the sea surface. After the earthquake, I felt the land slowly sinking, which seemed to continue for about 30 minutes. The subsidence of the ground surface reached up to three meters in some places.

He drew a map of the sunken village as shown in Figure 1.5.7.

During the field investigation, I also found numerous sand boils caused by liquefaction in Narvacan and its surroundings, confirming that the significant subsidence of the ground was caused by liquefaction.

Figure 1.5.8 shows an aerial photo of the sunk land, which roughly coincides with the map that the village chief wrote. The subsided area was shown by the pilot of a helicopter.

Figure 1.5.9 shows evidence of the settlement of the ground, where, according to villagers, "the water level rises up to the eaves during high tide as shown in the photo." Although the village is described as "submerged," the high elevation areas remain, and most residences are above sea level at low tide. The term "submerged" might give the impression that the entire village was instantly submerged underwater, but a more accurate description would be "the ground experienced significant subsidence due to liquefaction." Of course, living in the village became impossible, and the residents relocated to the mainland opposite to the bay, abandoning the village.

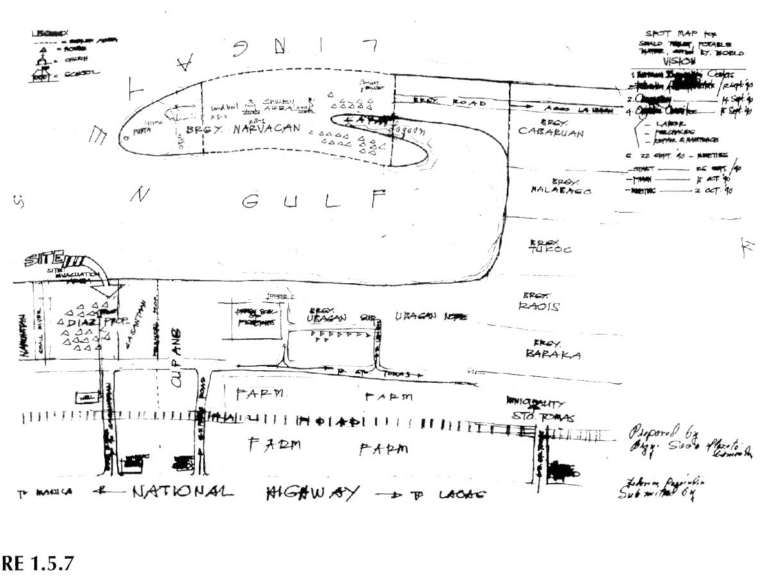

FIGURE 1.5.7
Map drawn by the village chief.

FIGURE 1.5.8
Aerial view of Narvacan village after the earthquake (aerial photo taken after sinking of the island).[16]

FIGURE 1.5.9
Situation of Narvacan village during low tide.

Figure 1.5.10 shows a large sand volcano found around the village, which suggests that extensive soil liquefaction occurred in the neighborhood.

FIGURE 1.5.10
A large sand volcano (1990 Luzon earthquake, Philippines).[16]

The author speculates that the situation of Uryu Island after the 1596 earthquake in Japan might have been like that of Narvacan village in the Philippines. The impossibility of habitation and the leaving of residents could lead to a similar fate as Uryu Island, eventually submerging under the effects of high tides and waves over many years. If we consider that some part of the island caused ground sliding, and another part subsided due to liquefaction caused by the earthquake and gradually sank, it might explain why there is no written history of Uryu Island's submersion for 100 years after the event. The 1990 earthquake in the Philippines might have helped to solve the mystery of Uryu Island's submersion that occurred over 400 years ago in Japan.

There are legends about islands in Japan that were submerged due to earthquakes or tsunamis, such as Kamoshima off the coast along the Japan Sea of Masuda City in Shimane Prefecture, believed to be the final resting place of a famous poet Kakinomoto Hitomaro (662–710) in the Heian era, Tosa-Minato in the Tsugaru Peninsula of Aomori, and Imakire in Lake Hamana in Aich Prefecture.

Similar events of sinking of the coast of Lake Sapanca during the 1990 Turkey Kocaeli earthquake have been reported, due to horizontal movement and subsidence of the ground. A hotel building in Figure 1.5.11 was moved toward the lake and tiered.

It is difficult to imagine that rock islands could suddenly sink tens of meters deep into the sea due to any major earthquake. It is presumed that these islands are submerged due to sliding and significant subsidence of soft ground caused by liquefaction.

FIGURE 1.5.11
Hotel building moved several meters horizontally due to liquefication-induced ground displacement (1999 Kocaeli earthquake in Turkey; see Figure 1.4.14).[14]

2

Measures for Mitigation of Future Disasters

2.1 Measures for Urban Infrastructure

2.1.1 Earthquake Risk of Coastal Industrial Parks

During the 2011 East Japan earthquake, 17 liquefied natural gas (LNG) tanks constructed on a reclaimed land of the Tokyo Bay exploded into huge fires. Fragments of the exploded tanks, 80 cm in width and 1.5 m in length at maximum, were scattered and fell in residential areas, although wide roads have been constructed between the industrial complexes and the residential areas, designed to block the effects of fires from the oil complexes (Figure 2.1.1).[17,18]

Furthermore, the earthquake caused extensive soil liquefaction in reclaimed lands along the Tokyo Bay, leading to many houses and buildings tilting and sinking. It also caused severe damage to lifelines, such as water, sewage, gas, and electricity, resulting in serious effects to the lives of residents (Figure 2.1.2).

FIGURE 2.1.1
Explosions of LNG tanks in Tokyo Bay (2011 East Japan earthquake).

DOI: 10.1201/9781003728788-2

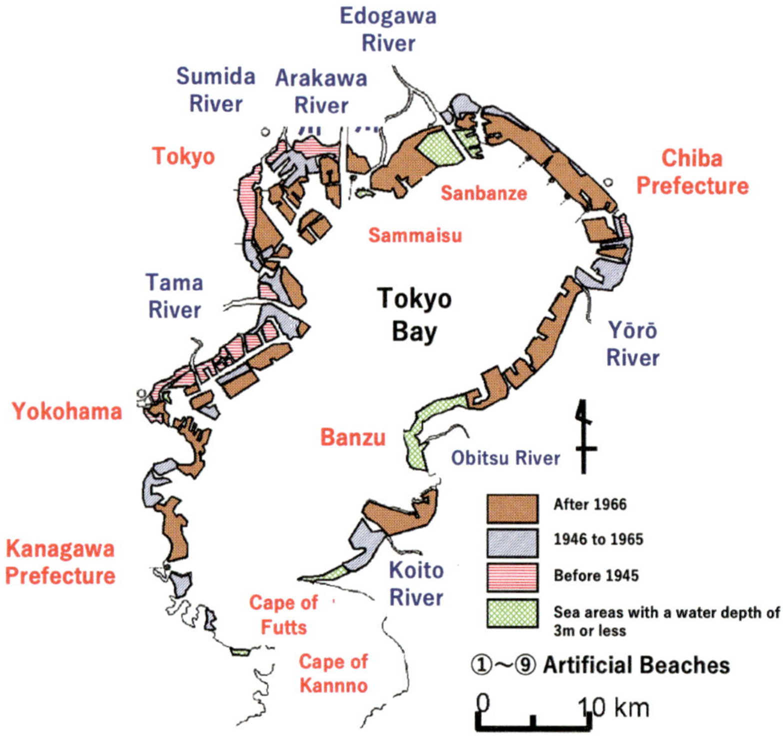

FIGURE 2.1.2
Reclaimed land of Tokyo Bay and its reclamation history (supplemented based on the compilation by Sōhei Kaizuka.)

Reclamation in coastal areas like Tokyo Bay and Osaka Bay has been carried out since the 15th century. After World War II, large-scale reclamation projects were undertaken for the revival of the country, leading to the construction of industrial parks. For example, in Tokyo Bay, 14 thermal power plants, three city gas plants, three steel mills, and eight oil refineries were built, forming the foundation for Japan's economic development. For the convenience of importing energy resources and raw materials such as oil, gas, and iron ore, and exporting manufactured goods, most industrial facilities have been constructed in coastal areas.

Furthermore, soil liquefaction caused serious damage to residential areas around the Tokyo Bay during the East Japan earthquake. A number of houses and buildings were sunk and tilted, while the heavy damage to lifelines made the residents suffer for a long time (Figures 2.1.3 and 2.1.4).

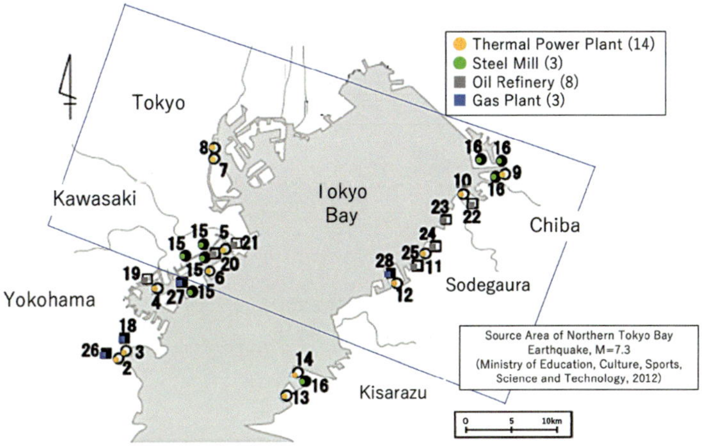

FIGURE 2.1.3
Major industrial complexes in the coastal area of the Tokyo Bay.

FIGURE 2.1.4
Liquefaction-caused damage: raising of sewer manhole in reclaimed area (2011 East Japan earthquake).

FIGURE 2.1.5
Fire of an oil refinery at Sendai Port (2011 East Japan earthquake).

These coastal reclaimed lands face increased earthquake risks due to the following factors:

1. Amplification of earthquake motion by the soft soil of reclaimed lands increases the seismic force to structures beyond design accelerations, leading to structural damage.
2. Liquefaction of reclaimed grounds, which were constructed by dredging seabed sediments or soils from nearby hills, consists of liquefiable soil.

Liquefaction-induced ground displacement is another factor that increases earthquake risks in reclaimed lands. Seawalls may move significantly toward the sea due to seismic force and soil liquefaction, causing horizontal displacement of the reclaimed land toward the sea. During the 1995 Kobe earthquake, seawalls in reclaimed lands of Kobe and Osaka displaced up to 7 m horizontally, damaging buried lifeline pipes and sewage treatment facilities.

In the 2011 East Japan earthquake, the tsunami caused extensive damage to coastal industries. A large fire broke out at an oil refinery at Sendai Port of Miyagi, shown in Figure 2.1.5, believed to be caused by a tank lorry swept away by the tsunami colliding with the refinery's pipeline. Since all employees had evacuated from the refinery at the time of the attack of the tsunami, the direct cause of the ignition was not identified.

Methods to protect structures against liquefaction and its induced ground displacement are discussed in Sections 2.2, 4.6, and 4.7.

2.1.2 Measures for Oil Tanks against Long-Period Earthquake Ground Motions[18]

Earthquake ground motions of several seconds dominant periods have caused oil tank fires during past earthquakes in Japan, such as 1964 Niigata, 1983 Japan Sea, and 2003 Off-Tokachi in Hokkaido earthquakes, as well as in the world, such as 1064 Alaska, 1999 Taiwan-Chichi, and 1999 Turkey Kocaeli earthquakes (Figure 2.1.6).

A typical example of long-period earthquake ground motions is shown in Figure 2.1.7, which were recorded at two locations, Tomakomai city and Hiro-o town in Hokkaido during Off-Tokachi earthquake. Two ground motions are very different from each other, even when the earthquake is the same. The ground motion observed in Tomakomai consisted of combined long-period motions of 6–8 seconds, particularly in the later part of the ground motion. On the contrary, the ground motion observed at Hiroo consisted of short-period motions of under 1.0 second. This difference resulted from the distinction of the ground conditions at the two locations. The ground at Hiroo consists of hard rock under a few meters of surface soil, while the ground at Tomakomai consists of soft sediment soil of 1–1.5 km thickness. The long-period ground motion induced the sloshing vibration of the oil content together with the floating roofs, resulting in big fires due to the metal collision between the steel floating roof and the steel side wall.

FIGURE 2.1.6
Fire of oil tanks (2003 Off-Tokachi earthquake, Tomakomai, Hokkaido).

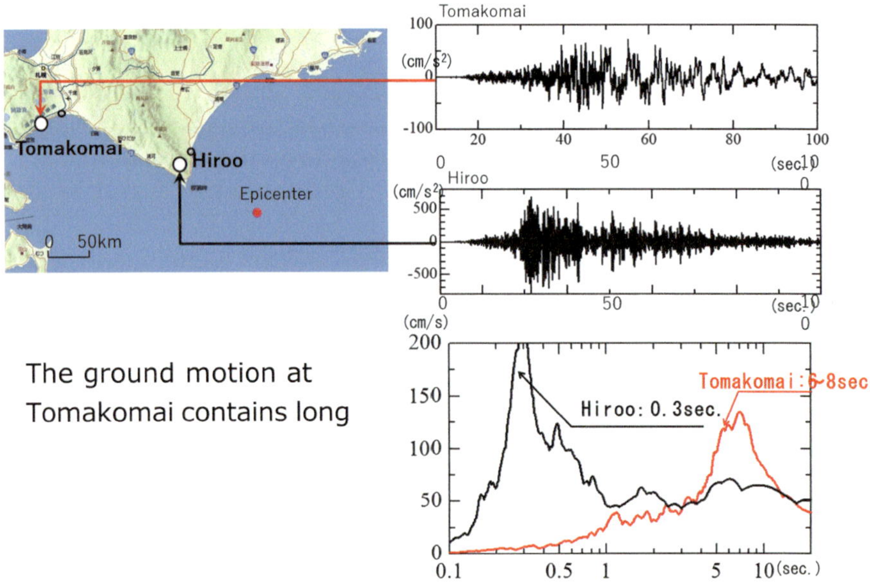

FIGURE 2.1.7
Long-period ground motion observed at two locations, Tomakomai and Hiroo (2003 Off-Tokachi earthquake, Hokkaido).

The natural vibration periods of the sloshing vibration of the oil in cylindrical tanks depend on the diameter of the tank and the depth of the oil. The natural periods of the fired tanks at Tomakomai were estimated as 7–8 seconds from the diameter of the tank and the depth of the oil, which mostly considered the dominant periods of the ground motion at Tomakomai, resulting in extensively high sloshing vibration (Figure 2.1.8).

Around the Tokyo Bay, more than 600 floating-roof-type tanks have been constructed in two districts, the Tokyo-Yokohama area and the Tokyo-Chiba area.

Under the assumption of a simultaneous occurrence of Tokai and Tonankai earthquakes along the Nankai Sea Trough in the Pacific Ocean (Figure 1.3.2), the long-period ground motions at Tokyo-Yokohama and Tokyo-Chiba districts were estimated as shown in Figure 2.1.9.

The rise of the oil surface of about 600 tanks was estimated by the following procedure. The depth of oil in all cylindrical tanks in the whole area of Tokyo Bay is measured simultaneously by using aerial photographs taken at the same time. The probabilistic distribution of the ratios of the depth to the height of the side walls is assumed not to be changed for the calculation. As a result of the calculation, it was estimated that the content oil spills out from about one-tenth of the total number of the flooding roof-type tanks (64 among 616 tanks).

How to keep away flowing out of a large amount of oil due to the sloshing vibration is one of the important subjects for the prevention of big fires

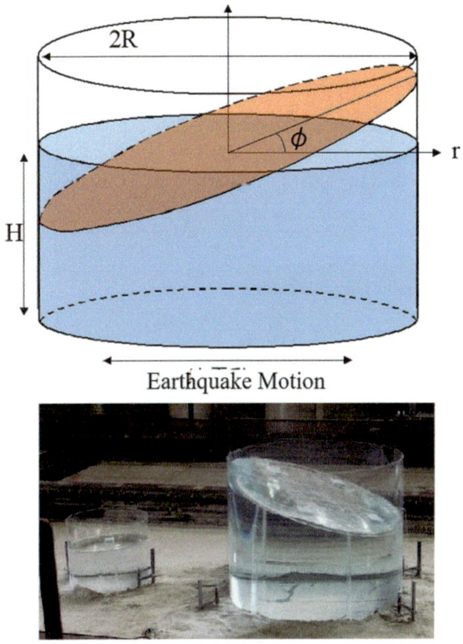

H: Height of Content Water
$2R$: Diameter of Tank
G: Gravity Acceleration
T: First period of Sloshing Vibration

$$T = 2\pi\sqrt{\frac{2R \times \coth(3.68 \times H / 2R)}{3.68 \times g}}$$

Diameter of Tank 2R (m)	High + Might of H (m)	Natural Period of Sloshing Vibration T1 (sec)
30	10	6.25
30	20	5.77
50	20	7.80
50	30	7.49
80	20	10.98
80	30	9.97

FIGURE 2.1.8
Natural periods of sloshing vibration of content in cylindrical tanks.

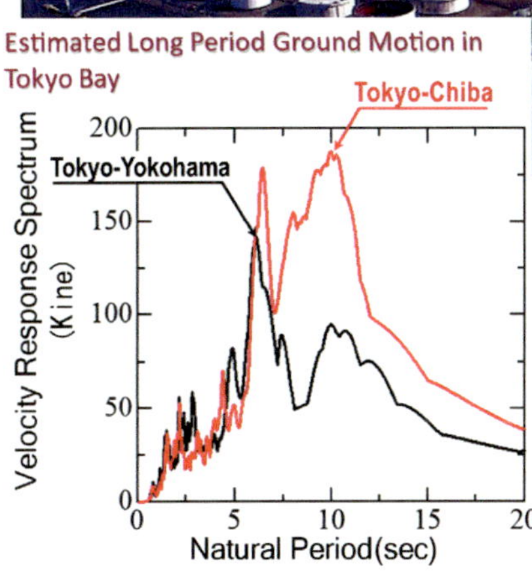

Diameter of Tank	Total number of Tank	Number of Oil-Spilled out Tanks
~24m	203	13(6.4%)
24~34m	136	27(19.9%)
34~60m	118	18(15.3%)
60m~	159	6(3.8%)
Total	616	64(10.4%)

FIGURE 2.1.9
Estimated earthquake ground motions (Tokai-Tonankai earthquake, Figure 1.3.2) and estimation of number of tanks of oil flowing-out in the Tokyo Bay.

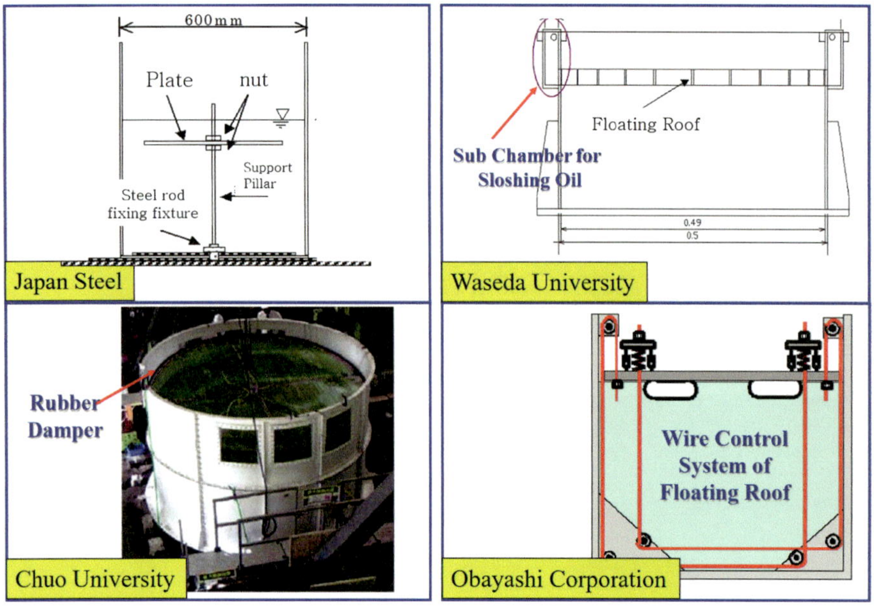

FIGURE 2.1.10
Proposals of measures for prevention of sloshing vibration of tank oil.

in industrial parks. A symposium was held in Tokyo in 2005 to discuss this problem. Several countermeasures for preventing the sloshing vibration of oil were proposed. Japan Steel Corp., Waseda University, Chuo University, and Obayashi Corporation proposed their ideas as shown in Figure 2.1.10, but at present, no idea has not been realized. There are several technical difficulties to stop or to reduce the sloshing vibration of oil. These measures should be mostly applied for existing large tanks of several 10m diameter which were fabricated from thin steel plates. The additional heavy devices cannot be installed on the existing tank walls.

At present in Japan, the following strategies have been recommended to prevent big fires at oil plants by the Fire Department of the Japanese Government (Figure 2.1.11):

1. Lowering the oil surface evenly by continuously adjusting the volume of oil in multiple tanks in a plant.
2. Reinforcement of the oil protective barrier walls, particularly the foundations against soil liquefaction; the barrier walls have been frequently damaged by soil liquefaction during past earthquakes.
3. Preparation of large-scale foam-type fire extinguisher.
4. Preparation of oil fence for the prevention of oil flow into the sea.

FIGURE 2.1.11
Rupture of concrete oil fence by soil liquefaction.

2.1.3 Seismic Design of Inground Tanks

In response to the 1973 oil crisis in the world, the importance of stockpiling energy resources such as oil and natural gas was seriously recognized in Japan, leading to the construction of above-ground and inground tanks, as well as rock cavern tanks. Regarding oil, a policy of the Japanese government to maintain a national stockpile equivalent to 60 days of consumption has been promoted and is still in place today.

The safety of the inground tanks during excavation works has been the most critical issue; if the excavation is safely conducted, it is believed that the structures would also be sufficiently safe against earthquakes. Therefore, seismic examination of inground tanks has yet to be performed. However, considering that the depth of inground pit can reach tens of meters and that the diameter of cylindrical tanks can exceed 50 m, verifying seismic safety became a crucial design issue for constructing large inground tanks in an earthquake-prone country like Japan.

While working at a construction company, the author was involved in a project funded by a research and development grant from the Ministry of International Trade and Industry (MITI) to study the seismic safety of large inground tanks for storing oil and LNG and to develop seismic design methods (Figure 2.1.12).

Earthquake observations were conducted on actual inground structures: a cylindrical tank excavated in reclaimed land and a rock cavern for oil stockpiling. The goal was to observe how these large inground structures deform during earthquakes and to develop seismic design methods based on the observed results.

FIGURE 2.1.12
LNG inground tanks in the reclaimed ground around the Tokyo Bay.

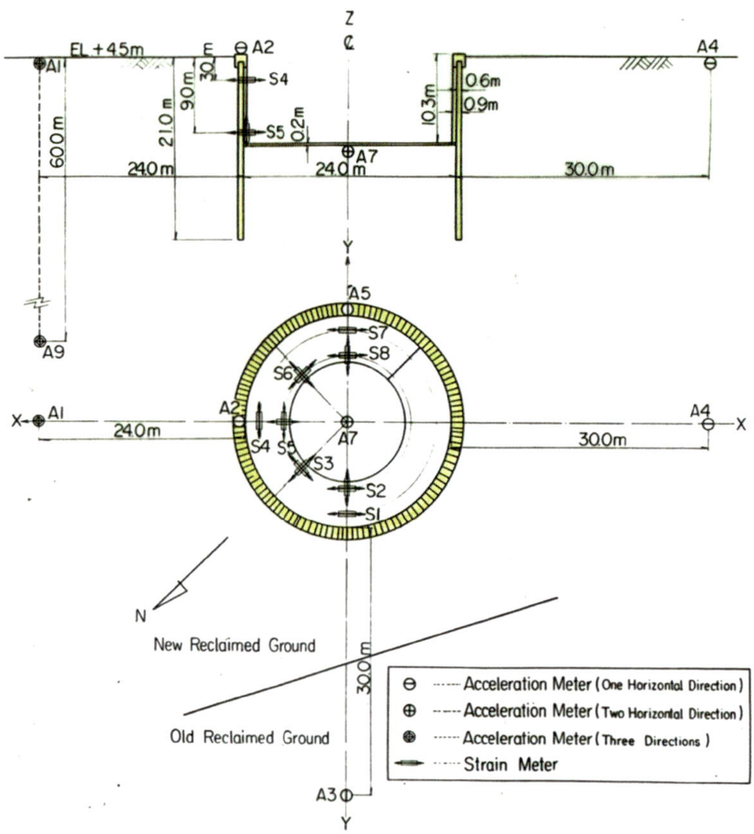

FIGURE 2.1.13
The cylindrical inground tank for earthquake observation.

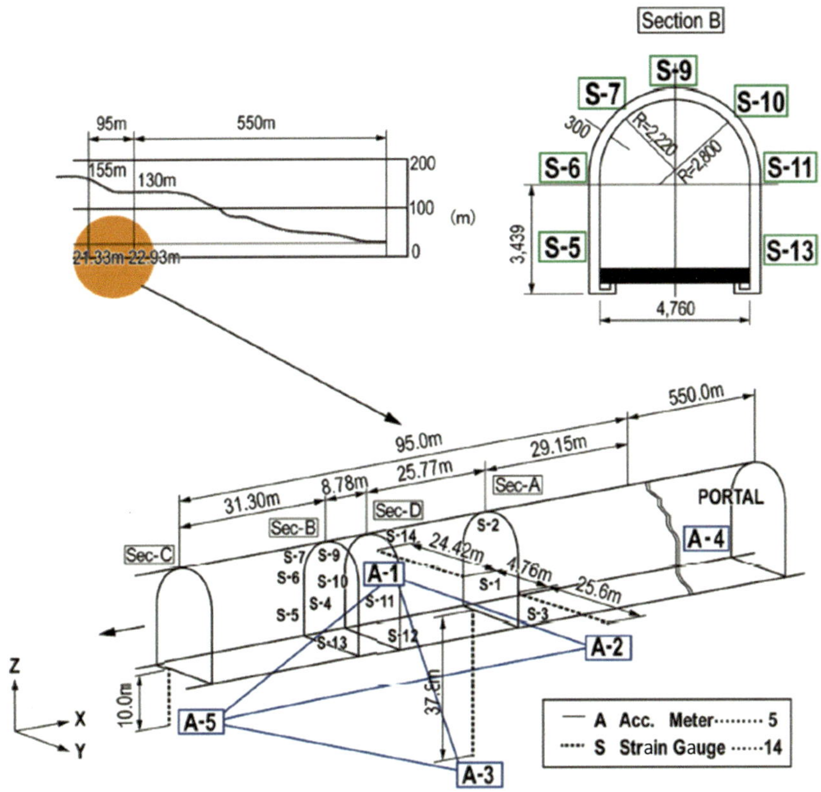

FIGURE 2.1.14
The tunnel for earthquake observation.

The cylindrical inground tank was a concrete tank in Shimizu City as shown in Figure 2.1.13, with a depth of 10 m and a diameter of 23 m. Figure 2.1.14 shows the rock cavern for a railway tunnel of the JR Sakari Line in Kameishi City of Iwate.

High-sensitivity strain gauges were installed to measure the strain on the tank sidewall and the tunnel inner surface to determine how the tank and tunnel would deform during earthquakes. Accelerometers were also placed around the surrounding ground.

Approximately one month after installing the measuring instruments in the Kamaishi tunnel, a magnitude 7.4 earthquake occurred off the coast of Miyagi in 1978, close to the observation site, shaking the Tohoku region severely. It was expected to provide valuable data. After the earthquake, the JR Tohoku Line was suspended, so my colleagues and I had to drive to Kamaishi. It took over 10 hours to reach, navigating through damaged and blocked roads. Expecting to find valuable observational data, we entered the measurement room in the tunnel, only to find no signs of the recorder having operated. The power was off, likely due to a widespread power outage caused by the earthquake in the Tohoku region. Unfortunately, no data from the Miyagi off coast earthquake was collected. In hindsight, an emergency power supply should have been prepared. Despite ongoing aftershocks, we managed to collect data by urgently procuring batteries and bringing them into the measurement room.

What did the earthquake observations reveal? It was found that the cylindrical underground tank deformed elliptically within the horizontal plane as shown in Figure 2.1.15, and the arch part of the railway tunnel also deformed into an elliptical shape as shown in Figure 2.1.16. The cause of the deformation of the inground tank and the tunnel was the relative displacement, the strain of the surrounding ground due to the propagation of seismic waves.

It is quite natural that a circular structure would deform into an elliptical shape when distorted, as the most likely deformation for a circular structure. A significant amount of time and research funding was invested to understand this simple phenomenon.

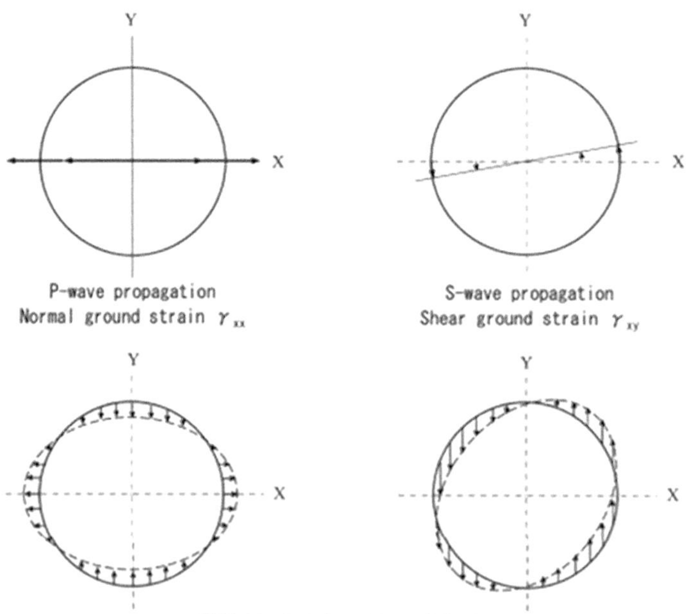

FIGURE 2.1.15
Deformation of cylindrical tank due to seismic wave propagation.

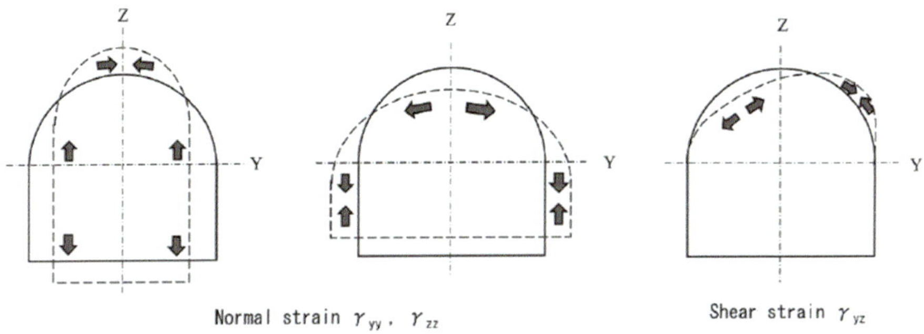

FIGURE 2.1.16
Deformation of cross-section of tunnel due to seismic strain of surrounding ground.

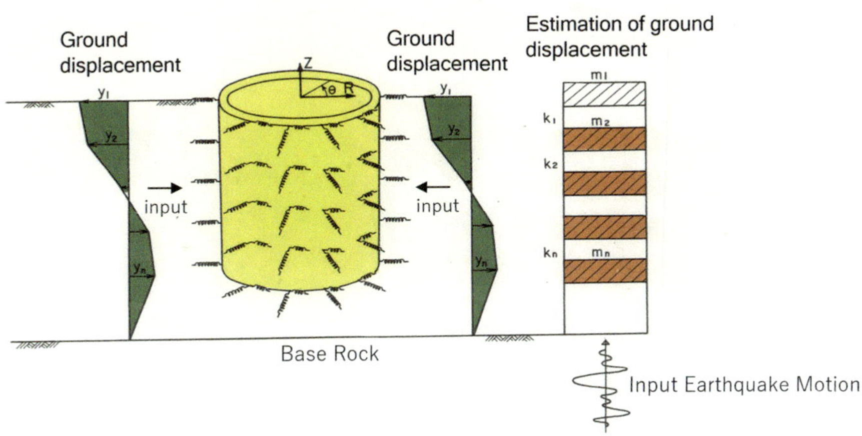

FIGURE 2.1.17
Seismic design of underground tank.

For the seismic design of underground tanks, a numerical method, shown in Figure 2.1.17, was proposed, where the circular tank is modeled as a cylindrical structure supported by a soil spring, and the ground displacement is input as forced displacement into the tank.

2.1.4 Seismic Design of Undersea Tunnels[19]

An undersea tunnel approximately 1 km in length has been constructed under the first navigational route of the Tokyo Port, between Oi in Shinagawa Ward and Daiba in Koto Ward (Figure 2.1.18). This tunnel is now a part of the Tokyo Bay Highway and serves as a crucial component of the metropolitan transportation network.

FIGURE 2.1.18
Undersea tunnel crossing the 1st. Route at Tokyo Port.

FIGURE 2.1.19
Submerged tunnel element constructed in dry dock.

The tunnel was constructed using the "submerging method." This construction method involves building the concrete tunnel boxes in a dry dock on land (Figure 2.1.19), towing them to the construction site, and then

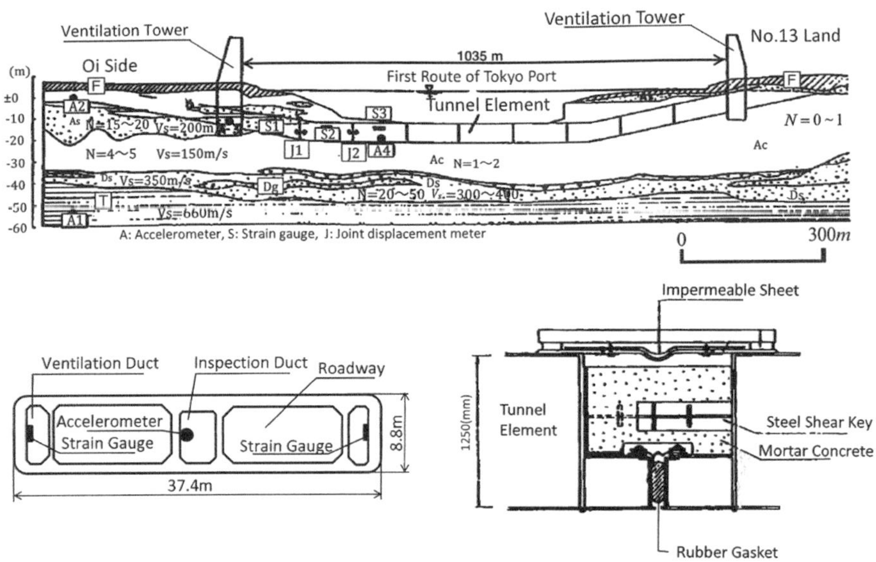

FIGURE 2.1.20
Structure of a submerged tunnel.

submerging them on the seabed to form the undersea tunnel. In the case of the first navigational route of Tokyo Port, each tunnel element is a concrete box structure, 110 m long, 37.4 m wide, and 8.8 m high, containing space for six car lanes. Then, such tunnel elements were submerged one after another on the seabed at the tunnel construction site and connected to form a tunnel of a length 1,035 m (Figure 2.1.20). This construction method has been used for many undersea tunnels in major metropolitan areas in Japan, serving as vital arterial roads.

However, the shield tunneling method was adopted for the Tokyo Bay Crossing Tunnel between Kanagawa and Chiba prefectures. The tunnel was excavated at a much deeper position from the seabed surface, ensuring safety against earthquakes. In contrast, the submerged method places the tunnel on the seabed surface, offering the advantage of shorter approach roads for entering the tunnel from both banks.

The buoyancy acting on the tunnel elements reduces the vertical force exerted on the seabed when the elements are placed on the seabed, allowing the construction even on soft and weak grounds. However, since the tunnel is installed shallowly beneath the seabed, its safety against earthquakes is one of the most concerned matters.

The submerging method was originally developed in Europe, where the seismic activity is very low, and design against earthquakes is not required. When introducing this method to Japan, an earthquake-prone country, concerns were raised about whether using buoyancy to land on the soft seabed would pose a seismic risk. Therefore, the development of design methods

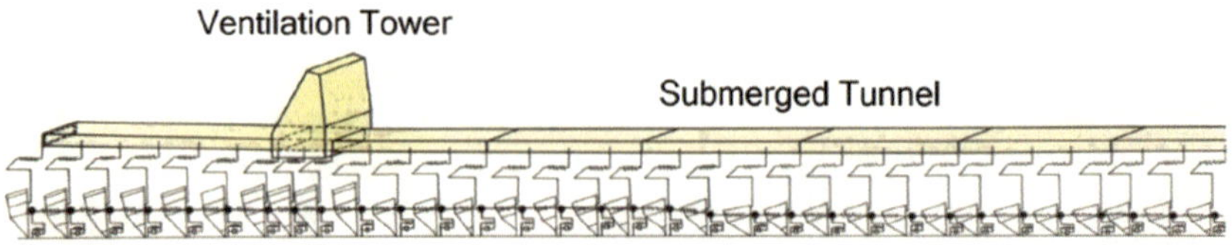

FIGURE 2.1.21
Seismic design model of submerged tunnel.

and analytical models for verifying the seismic safety of submerged tunnels became one of the important issues. Based on the results of shaking table tests with models integrating the ground and tunnel, a seismic design method for submerged tunnels was developed. This design method focuses on the axial forces and bending moments generated in the tunnel due to the differences in dynamic response between the ground on the land and the seabed. This earthquake-resistant design method is known as the "response displacement method."

The response displacement method has been used for the seismic design of linear inground structures like buried pipes. It is based on the concept that the deformation of inground structures during earthquakes is governed not by the acceleration of earthquake motion, but by relative displacements of the surrounding ground. While above-ground facilities and structures are designed based on inertial forces, i.e., mass multiplied by acceleration (the acceleration method), the term "displacement method" would have been more appropriate for this design approach. However, the "displacement method" was already used as a technical term in the field of structural mechanics at the time, so the "response displacement method" was chosen to include the meaning of calculating dynamic response ground displacements. It is now widely used for the seismic design of buried pipes, underground ducts, and inground tanks.

For the seismic design of a submerged tunnel, a numerical model is used as shown in Figure 2.1.21, where the ground displacements are forced to the tunnel through a soil spring. The relative ground displacements between both lands and seabed are the governing factor for the deformation of the tunnel, which are estimated by considering the difference of dynamic response of the surface ground between the land sides and the seabed.

Following the earthquake-resistant design of the first navigational route tunnel in Tokyo port, the response displacement method has been used for many other undersea tunnels. The construction of the Bosphorus Strait Tunnel in Istanbul, Turkey, was carried out at a water depth of 60 m, unprecedented for undersea tunnels in Japan. Despite various challenges, such as strong currents, the construction was successful under the leadership of Japanese engineers.

2.2 Measures against Liquefication and Its Induced Large Ground Displacements[20]

2.2.1 Prevention of Liquefaction-Induced Ground Displacements

As part of the National Land Resilience Strategy for critical infrastructures that began in 2014, seismic reinforcement of coastal industrial parks has been proceeding in Japan, and measures have been developed to prevent liquefaction and its induced large ground displacements. Countermeasures such as steel pipes or steel sheet pile walls construction behind existing quay walls and soil improvement methods behind quay walls have been developed and used to prevent liquefaction-induced ground displacements. However, for both the construction of inground walls by steel piles and soil improvement, a large construction cost is required, making it generally difficult for private companies. Although there are national subsidy programs for quay walls reinforcement, the extensive length of quays requiring reinforcement has yet to be widely implemented.

The research group at Waseda University developed the "skip-driven pile method," which prevents ground displacements by driving steel pipe piles in a single row or double rows at a proper interval behind the existing quay walls. The developed method can reduce the cost and the period for the construction (Figure 2.2.1).

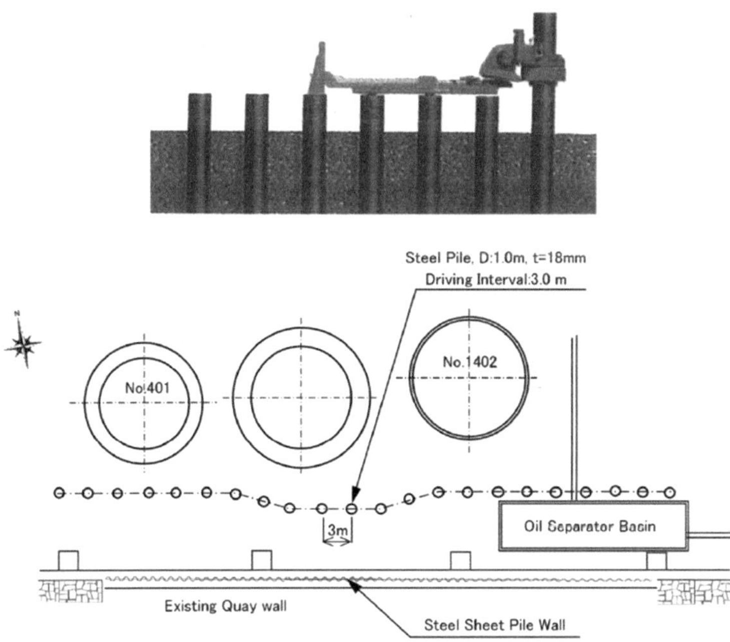

FIGURE 2.2.1
Protection of tanks against liquefaction-induced ground displacement by skip-driven piles.

The effectiveness of the method was verified by model experiments under a centrifugal loading condition as well as by two and three-dimensional finite element numerical analyses. This construction method has been implemented to prevent liquefaction-induced ground displacement at actual oil tank yards. Model experiments have shown that even with pile spacing 3 times the pile's outer diameter, sufficient reduction of ground displacements can be achieved.

The flow of liquefied soil stops when piles are driven in a row with a proper interval. Liquefied soil possesses both fluid and solid properties. The sand's properties change from liquid state to solid state resulting from the rapid expansion of the volume due to shear deformation of the soil. The sudden drop of pore water pressure ends the liquefaction state, restoring the rigidity of the soil.

The skip-driven pile method utilizes such properties of sand. Instead of blocking the movement of liquefied soil with a continuous wall, the method allows for controlled movement of sand by spacing the piles, which settles the liquefaction and, as a result, reduces ground displacement. Spacing the piles significantly reduces the cost of steel pipe materials, and the cost of pile driving is also reduced. The effectiveness of the skip-driven pile method has been confirmed by shaking table tests and 2D-3D numerical analysis. A similar construction method has been used for slope stability (Figure 2.2.2).

FIGURE 2.2.2
Prevention measures of liquefaction-induced ground displacement in a tank yard using skip-driven piles.

2.2.2 Protection of Residential Houses against Liquefaction

Every time a medium- to large-scale earthquake occurs, liquefaction has occured in reclaimed residential lands, alluvial grounds along rivers and lowlands, causing the subsidence and tilting of houses and buildings. Local governments have compiled liquefaction maps and made them public to raise awareness among residents about the likelihood of liquefaction. However, residential land development continues in grounds prone to liquefaction, often without any countermeasures in place.

While the inclination and subsidence of houses do not directly threaten the lives of residents, the psychological effects of living in such a house can significantly impact their post-earthquake life. As countermeasures against liquefaction for residential houses, solidifying the foundation ground with cement milk or additional foundation piles are recommended, but these methods are expensive and have yet to become widespread. The difficult challenge in liquefiable residential areas is the measures for existing houses. Even if liquefaction maps by municipalities indicate a high likelihood of liquefaction, implementing countermeasures for existing homes is generally difficult due to the cost and the space of construction. There are proposals to insert pipes into the liquefied layer beneath houses from its surroundings and inject solidifying materials like cement milk, but the increased construction costs remain a significant burden (Figure 2.2.3).

During the 2011 East Japan earthquake, large-scale liquefaction occurred in coastal reclaimed areas such as Urayasu City along the Tokyo Bay, causing

FIGURE 2.2.3
Tilted house by soil liquefaction (2011 East Japan earthquake).

many houses to tilt or sink. After the earthquake, one proposed method to protect residential areas from liquefaction was to encircle entire blocks prone to liquefaction with impermeable walls and pump up the groundwater surrounded by these walls. This method aims to prevent liquefaction by removing porewater between sand particles, unsaturated state. This block-wide approach could reduce the construction costs per house, but it requires consensus among all residents, and there are ongoing issues, such as the need to pump groundwater up and the associated costs continuously (see Section 4.6.2).

I have also been involved in liquefaction countermeasures by groundwater pumping at an oil refinery in a Kawasaki City industrial complex during my work for a construction company (see Section 4.6.1). Thirty years later, a re-investigation of the effectiveness of these liquefaction countermeasures revealed that the pipes of the groundwater pumping wells were clogged, and the groundwater level had stayed high elevation. The author's research group proposed a method for liquefaction countermeasures for existing houses. This method is to enclose the foundation ground with lightweight steel sheet piles. This method does not involve pumping groundwater. By constructing lightweight steel sheet pile walls to prevent the horizontal movement of liquefied soil, the method aims to prevent the subsidence and tilting of buildings. The idea is that if the foundation ground liquefies but the liquefied soil under the basement of houses cannot be pushed out horizontally, subsidence and tilting can be prevented.

The effectiveness of this method was confirmed by model experiments under centrifuge conditions. This method was planned for houses in Urayasu City, where many houses were sunk and tilted during the 2011 East Japan earthquake, but it was not realized because the individual cost burden exceeded personal limits. According to cost estimates for a standard-sized home, direct construction costs such as materials and installation of sheet piles could be kept below 50,000 U.S. dollars. However, indirect construction costs, such as removing and restoring parts of roofs and concrete floors, which will be an obstacle for driving sheet piles and relocating and replanting vegetation in gardens, increased the total cost beyond 50,000 U.S. dollars. No residents were willing to bear such costs for liquefaction countermeasures.

Residents' financial situations vary, and the amount they can spend on liquefaction countermeasures for their current houses is generally limited to about 20,000–30,000 U.S. dollars. There is a need for more affordable liquefaction countermeasure methods, but these have yet to be realized.

Figure 2.2.4 is another proposal by the author's research group to reduce subsidence of houses by driving short sheet piles around the foundation of existing houses. Experiments on a shaking table proved the effectiveness of this method. The experiment showed that the subsidence of houses can be reduced to one half, if the sheet pile is driven to about half the depth of liquefied soil.

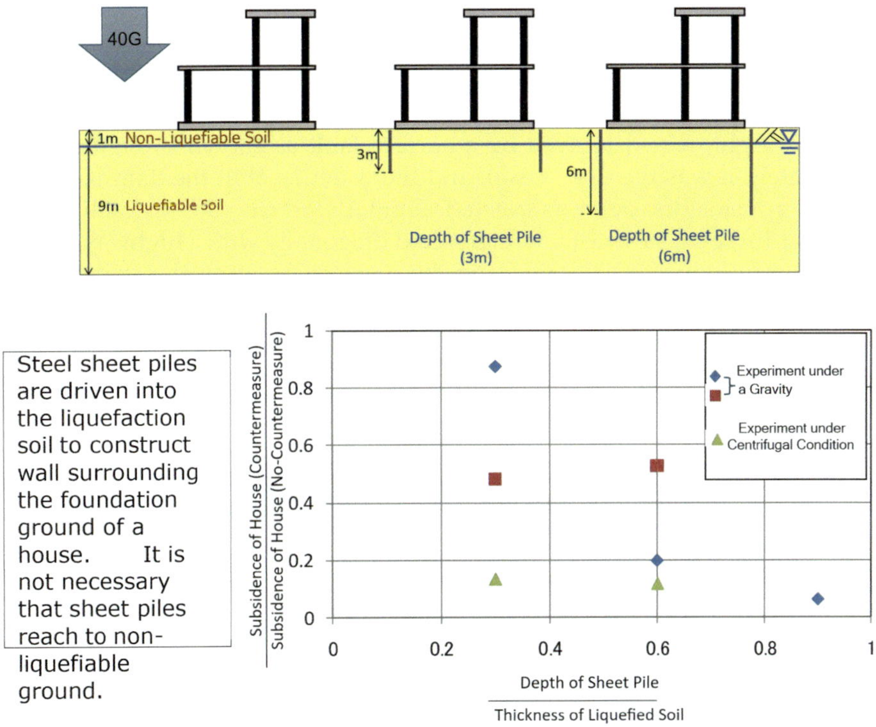

FIGURE 2.2.4
Liquefaction countermeasures for houses (model of an experiment under centrifugal loading condition).

2.3 Measures against Ground Ruptures

2.3.1 Surface Fault-Induced Ground Ruptures

Faults that cause earthquakes are referred to as "active faults." When the rupture caused by an active fault propagates upward and appears on the ground surface, it is called a "surface fault." In Japan, there have been a few cases where major infrastructure facilities have been damaged by surface faults. During the 1923 Kanto earthquake, the Tanna Tunnel of the Tokaido Railway Line between Tokyo and Osaka, under excavation, was crossed by a surface fault. However, the tunnel was completed by re-excavating the fault crossing zone.

In the 1995 Kobe earthquake, the distance between the two main towers of the Honshu-Shikoku connecting suspension bridge of a length 1,991 m, under construction, stretched by 80 cm due to fault displacement, but since the bridge girders had not been suspended yet, substantial damage was avoided.

After the earthquake, the length of the bridge girders was extended, and the bridge was successfully completed.

Internationally, there have been reports of damage caused by surface faults. In the 1999 Chi-Chi earthquake of Taiwan, a concrete dam for water storage, 25 m in height, was destroyed by a surface fault as shown in Figure 2.3.1. Fortunately, it was the dry season, and the water level in the dam lake was low, so a major disaster was averted. Similarly, in the 1999 Kocaeli earthquake in Turkey, a surface fault intersected diagonally with a highway bridge as shown in Figure 2.3.2, increasing the distance between the bridge piers and causing the girders to fall. Fortunately, no lives were lost.

FIGURE 2.3.1
Damaged concrete dam by surface fault (1999 Taiwan Chi-Chi earthquake).

FIGURE 2.3.2
Collapse of highway bridge by surface fault (1999 Turkey Kocaeli earthquake).

Expansion of the beam seats to prevent bridge collapse

Restoration Structural Members (Upper Chord and Shoes)

FIGURE 2.3.3
The fault countermeasures for the Tokaido Shinkansen Line.

There are examples of countermeasures being taken against surface fault movement. In Japan, when the Tokaido Shinkansen (the bullet train) was being constructed, it was predicted that the Iriyamase Fault on the right bank of the Fuji River in Shizuoka would move in conjunction with the anticipated Tokai earthquake along the Pacific coast of the main island of Japan. To prevent the bridge girders from falling due to fault displacements, the width of the girder seats (the top part of the bridge pier where the bridge girder is placed) was expanded as shown in Figure 2.3.3. Additionally, structural members of the bridge that were predicted to be damaged by the fault displacements were identified, and spare members were stored in a warehouse built on the right bank of the river. Fortunately, to date, there has been no earthquake that required the use of these spare members. Storing spare structural members in the warehouse is a measure aimed at the rapid restoration of the Shinkansen in case of damaging earthquakes. Structurally protecting buildings and facilities against surface faults is practically impossible.

During the excavation of the foundation rock for a 102 m high concrete gravity dam built on New Zealand's South Island, a fault was discovered directly beneath the dam foundation. Although relocating the dam was considered, no suitable site was found nearby, and it was decided to continue construction at the planned location. A design was adopted that included slits (concrete separation) in the concrete dam body to absorb fault displacement through relative movement of the dam body between slits as shown in Figure 2.3.4.

Fortunately, there has yet to be an earthquake to test the effectiveness of the measure. When a fault propagates toward the surface, it is known that the fracture surface can branch into multiple fractures or that the angle of the fracture surface relative to its vertical axis can change depending on the composition of the surface layers. Even if the position of the fault at depth is accurately estimated, it is difficult to reliably predict the location of the fault rupture that appears at the surface.

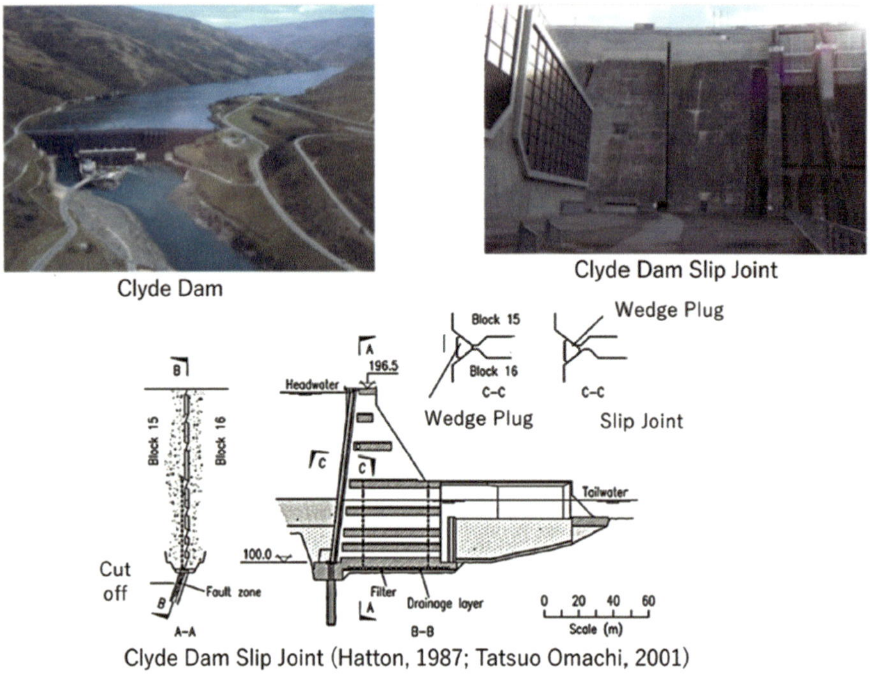

FIGURE 2.3.4
Measure for gravity dam against active fault.

2.3.2 Collapse of Underground Caverns

In the summer of 1900, a call came to my laboratory from Mr. Yanagawa, the mayor of Mitake Town in Gifu Prefecture. Before being elected mayor in 1995, Mr. Yanagawa had served as a disaster prevention commentator for NHK (Japan Broadcasting Association) for many years, and he was an acquaintance from that time.

Mayor Yanagawa's request was as follows (Figure 2.3.5):

> In Mitake Town, lignite had been mined before World War II. Lignite is a mineral fuel that is a precursor to coal, and though it has a lower calorific value than coal, it was widely used during times of fuel shortage before and after World War II. The underground cavities of abandoned mines after lignite extraction have mostly been left unfilled. These cavities, including those under urban areas, are widespread throughout the town, and occasional collapses have caused damage to buildings and houses. Mitake Town is close to the predicted source areas of the Tokai and Tonankai earthquakes of a magnitude 8.5 along the Nankai Sea Trough in the Pacific Ocean and there is a fear that the entire town could collapse if these earthquakes occur. However, the exact locations and depths of these underground cavities are not well understood. Could you investigate this troublesome problem at Waseda University?

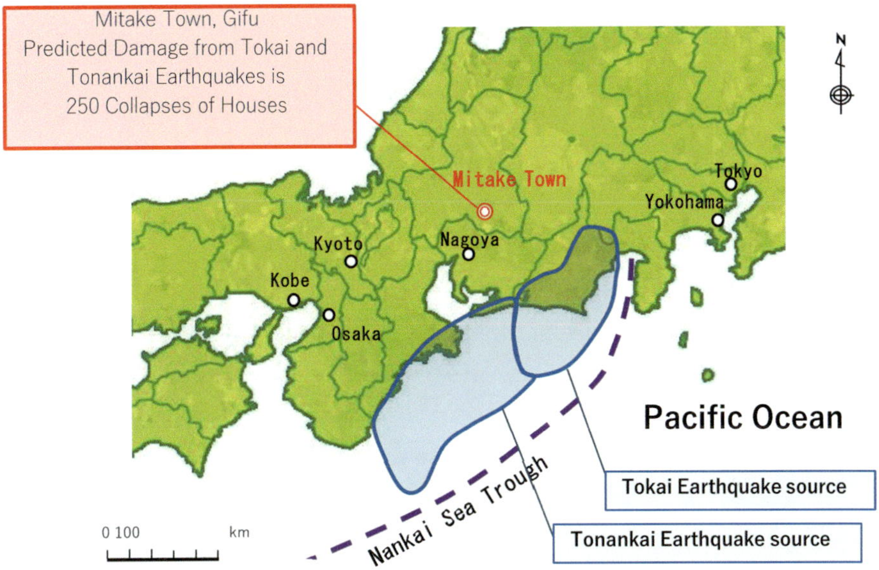

FIGURE 2.3.5
Mitake Town and predicted source area of Tokai and Tonankai earthquakes.

Given the potential impact on the lives of the town's residents and the request from an old acquaintance, I agreed to take on the project.

First, it was necessary to clarify the distribution of the cavities' locations and depths throughout the town. To this end, geological data were collected to estimate the distribution of lignite layers, as well as abandoning mine maps and boring data compiled by the former Ministry of International Trade and Industry, and created a three-dimensional distribution map of the lignite layers, as shown in Figure 2.3.6.

Various methods were tried to explore the depths and extents of the underground cavities from the surface, including (i) acoustic exploration (sending sound waves from one borehole and receiving them at another borehole to investigate the distribution of geological layers), (ii) surface-wave exploration (generating waves by striking the ground surface with a hammer-like object and estimating the depths and distribution of cavities based on how the waves propagate), and (iii) gravity change exploration (focusing on the slight reduction in earth's gravity due to underground cavities). A research group from the University of Miami, U.S.A., also participated in the investigation using acoustic tomography, a type of acoustic exploration used by the U.S. Navy to investigate seabed soil quality worldwide.

However, it became clear that the depths and the locations of the cavities could only be reliably explored with one method. The most reliable method to identify the depths and locations was interpolation though multiple boreholes.

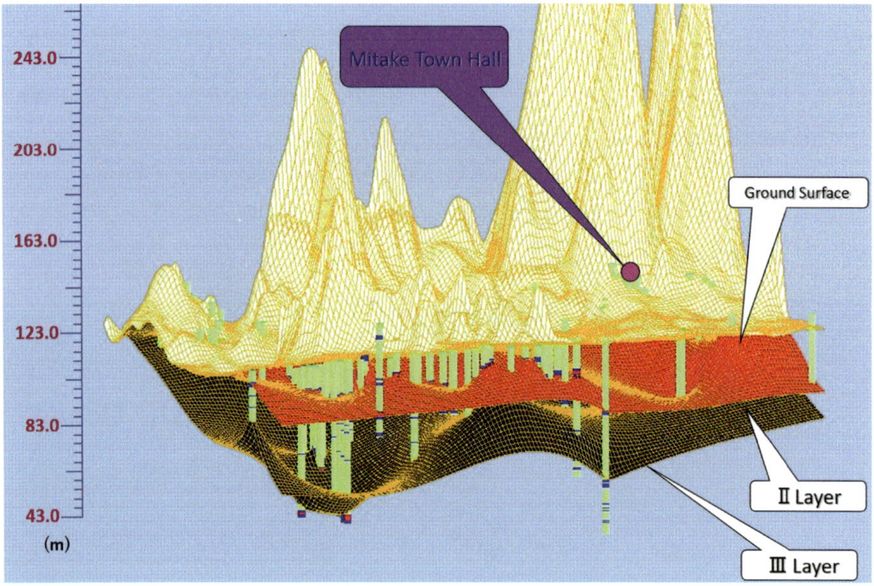

FIGURE 2.3.6
Three-dimensional distribution of lignite layer in Mitake Town.

To estimate the overall location and depth of the cavities, existing borehole data for building constructions, road works, and laying water and sewage pipes within the town were collected.

By using these data, three-dimensional soil strata map was made, which to whole area of the town through two layers II and III with the depth from less than 15 m to more than 30 m (Figure 2.3.6). The collapses of shallow cavern has damaged residential houses and lifelines like road, water, and sewage.

The survey by Waseda University encountered the reluctance of many residents to receive the investigation. They said that "everyone was well aware that there were cavities in the center of the town, and re-investigating and publicizing this under the mayor's initiative would not solve anything." Additionally, "the relatives of current residents, such as their fathers and grandfathers, who actually worked in the lignite mines and left the cavities unfilled, cannot be held accountable by the government or local authorities." and "Any further investigation is a nuisance." Mayor Yanagawa himself was originally from somewhere other than Mitake Town. He had been elected mayor as a so-called parachute candidate for the mayor election (Figure 2.3.7).

For the understanding and support from residents to the investigation, it was necessary to explain the scale and location of cavities that could collapse due to the predicted Tokai and Tonankai earthquakes, as well as the potential damage to houses and lifelines caused by such collapses. It was also necessary to lobby the national government and Gifu prefectural government for funding and support for filling the cavities, with Waseda University assisting in compiling explanatory materials for residents and presenting at community meetings.

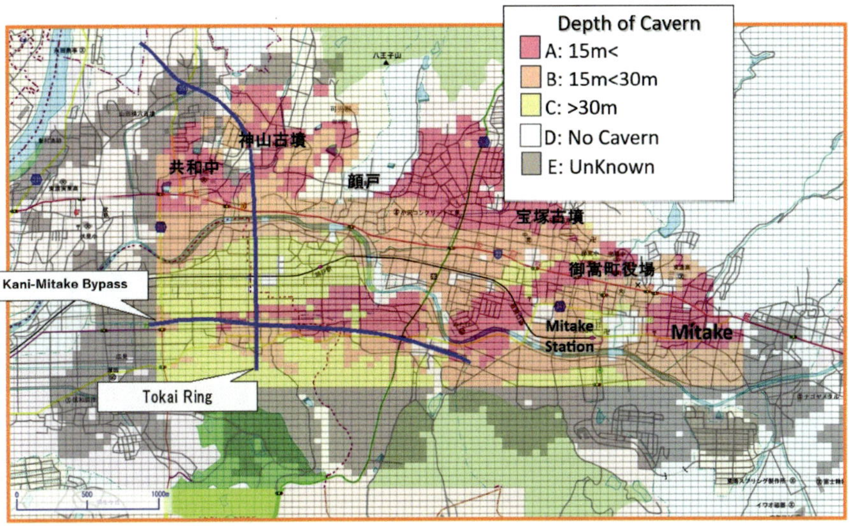

FIGURE 2.3.7
Estimation of location and depth of lignite cavern.

Given the vast amount of cavity that needed filling throughout the town, it was crucial to minimize construction costs. It was also required to fill the cavities completely without affecting surface structures or inground constructions, ensuring that the filling material reached the targeted spaces. Therefore, the fluidity of the filling material needed to be sufficiently maintained, and since the cavities were continuous without interruption, it was necessary to segment the filling spaces before filling. Consequently, a method known as the "limited filling method" was developed, led by two engineers Kenro Sugiura and Akio Sakamoto of Tobishima Construction company. Mr. Sakamoto was awarded a Doctor of Engineering Degree by Waseda University for developing this method and applying it in practice.

The risk of cavity collapse has also become a social issue in the cavities left by Oya stone mining in Utsunomiya City, Tochigi Prefecture. In the case of Oya stone, the large scale of the cavities raises concerns about increased construction costs, and a clear solution has yet to be proposed.

The cavity filling in Mitake Town has progressed as part of the national government project "Enhancement for Disaster Resilience of National Land." The Agency for Natural Resources and Energy under the Ministry of Economy, Trade and Industry has provided a grant for filling lignite mines, and Gifu Prefecture has used this fund to cover two-thirds of the filling construction costs for Mitake Town. Initially, the author had thought filling the cavities in Mitake Town was impractical due to the wide area and substantial costs involved, but it became a reality with the tailwind of the National Land Resilience Policy.

The future challenge is how to expand the investigation and filling operations conducted in Mitake Town to surrounding municipalities.

Many municipalities with lignite mines beneath residential areas remain, and the opinion that "we don't want any unnecessary actions" is still strong in these areas. However, the filling in Mitake Town has undoubtedly increased safety against future Tokai and Tonankai earthquakes, and the residents now seem to understand this well. The filling in Mitake Town will serve as a catalyst for nationwide cavity-filling efforts (Figure 2.3.8 and 2.3.9).

FIGURE 2.3.8
Collapse of abandoned mine.

FIGURE 2.3.9
Abandoned lignite cavern.

3

Global Increase of Natural Disasters and International Corporations

3.1 Global Increase of Natural Disasters

3.1.1 Weather Disasters

Weather disasters such as storms, heavy rains, floods, heat waves, forest fires, and droughts are increasing in the world. Figure 3.1.1 shows the number of weather-related disasters worldwide that resulted in more than 1,000 casualties over 70 years, from 1951 to 2023, in five-year intervals. Since 1991, the frequency of weather-related disasters has increased, and this trend has continued into the 21st century. Particularly, it should be noted that during the three years since 2021, abnormally high temperatures due to heat waves have drastically increased in Africa and Europe. It is important to carefully watch whether this drastic climate change will persist in the future.

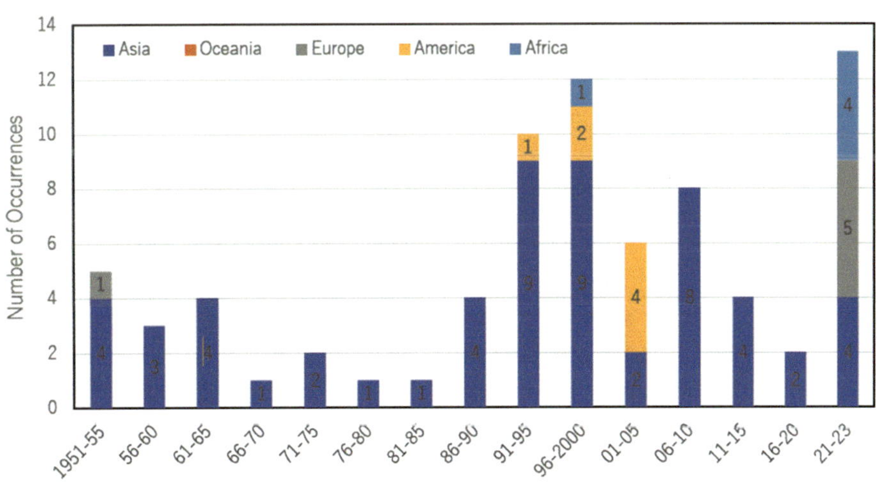

FIGURE 3.1.1
Occurrences of weather disasters (typhoons, heavy rains, floods, heat waves, etc.) that resulted in more than 1,000 casualties each (1951–2023, 75 occurrences in the world).

DOI: 10.1201/9781003728788-3

What is the cause of the increase in weather disasters? Global warming and the accompanying changes in the natural environment are believed to be contributing factors. The deterioration of the natural environment, such as sea-level rise due to temperature rises, urban heat island effects, loss of forests and farmland, desertification, and erosion of rivers and coastlines, is thought to increase the occurrence of weather disasters.

Over about 70 years (1951–2023), there have been 75 weather disasters worldwide that caused more than 1,000 casualties each, with 58 of these occurring in Asia. During the same period, the global death toll by weather disasters was 2.92 million, with approximately 97% (2.86 million) of these casualties occurring in Asia.

3.1.2 Earthquake and Tsunami Disasters

As shown in Figure 3.1.2, earthquake and tsunami disasters have also increased since 1986, although there has been a decrease in the last decade (2011–2020).

Figure 3.1.3 shows the number of earthquakes worldwide with a magnitude of 7 or higher. Although there has been a slight increase in the number of earthquake and tsunami disasters since 1991, this does not correspond with the rapid increase in the number of earthquake and tsunami disasters, shown in Figure 3.1.2. During the 70 years from 1951 to 2020, there have been 45 earthquake and tsunami disasters worldwide that resulted in more than 1,000 casualties each, with 31 of these occurring in Asia. During the same period, the global death toll from earthquakes and tsunamis was 1.37 million, with 69%, 0.88 million, of these casualties in Asia. Changes in social

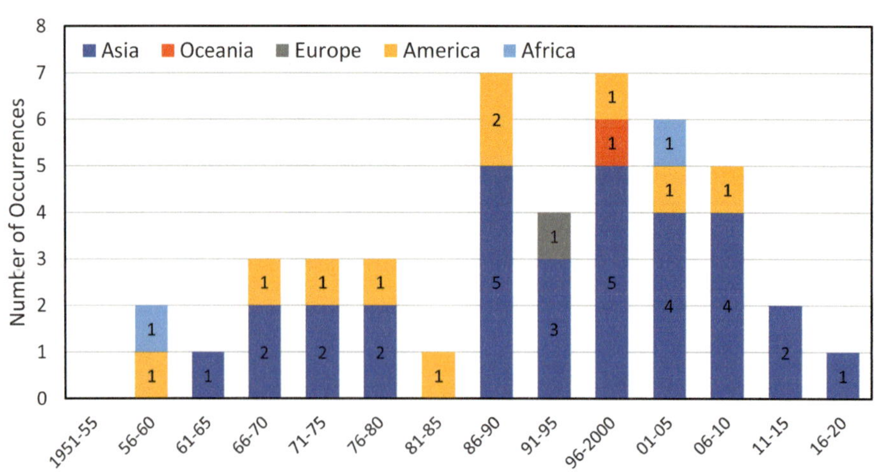

FIGURE 3.1.2
Occurrences of earthquakes and tsunami disasters those resulted in more than 1,000 casualties each (1951–2020, 45 occurrences in the world).

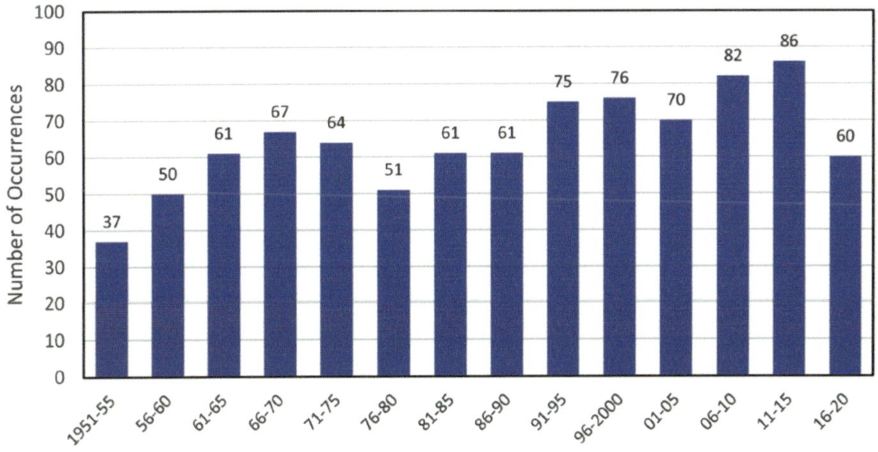

FIGURE 3.1.3
The number of earthquakes with a magnitude 7.0 or higher (1951–2020, 901 occurrences in the world).

structures, such as declining birthrates, aging populations, nuclear families households, urban overpopulations, and rural depopulation, are increasing vulnerability to disasters. Furthermore, changes in lifestyle, such as the decline in mutual aid awareness in urban areas, the lack of disaster experience transmission, and excessive dependence on electronic devices, are becoming risk factors for disasters.

In Asia, the collapse of houses with low earthquake resistance, such as those made of mud bricks, is the main cause of the loss of human lives. Due to financial issues, seismic reinforcement of houses has not progressed. Poverty exacerbates natural disasters, which in turn worsen the extent of poverty, creating a vicious cycle. Breaking this cycle requires more than just disaster prevention efforts; comprehensive support, including national land restructuring and social infrastructure development, is needed.

3.2 Strategies and Organizations for Natural Disaster Mitigation

3.2.1 Recommendations by the Science Council of Japan

In November 2000, the Science Council of Japan formed a special committee titled "International Cooperation for Natural Disaster Mitigation," and after over a year of discussions, the following recommendations were publicly announced. The author was engaged in compiling the recommendations. Here is a summary of the recommendations:

Current Status of International Cooperation in Disaster Mitigation by Japan: International support in the field of disaster prevention is seamlessly related to activities in society, economy, environment, education, etc., and close coordination between sectors is essential. However, international support in the field of disaster prevention by Japan has been carried out with insufficient coordination among relevant ministries and government agencies, universities, research institutions, and NPOs/NGOs.

In international cooperation for disaster mitigation, consistent and uninterrupted support from preventive measures to emergency response and recovery/reconstruction works are crucial. Coordination among responsible agencies at each stage needs to be improved, resulting in integrated international support.

Basic Policy for International Cooperation: There are high expectations, especially from Asian countries, for Japan, which has overcome diverse and severe disasters and achieved economic development. Meeting these expectations should be positioned as the basic strategy of Japan's international support for natural disaster mitigation.

Establishment of the "International Strategy Conference on Natural Disaster Mitigation": Based on the status and basic policies of international cooperation in disaster mitigation, the Science Council of Japan proposes the establishment of the "International Strategy Conference on Natural Disaster Mitigation." The conference will be responsible for "formulating an international strategy for natural disaster mitigation, "developing and leading an international framework of action," "organic integration of sectors," and "communication to society and consensus building among citizens." Under the conference, platforms for disaster prevention cooperation and support for affected areas, personnel training, and response to international programs will be established to coordinate institutions and organizations involved in international cooperation in the field of disaster mitigation (Figure 3.2.1).

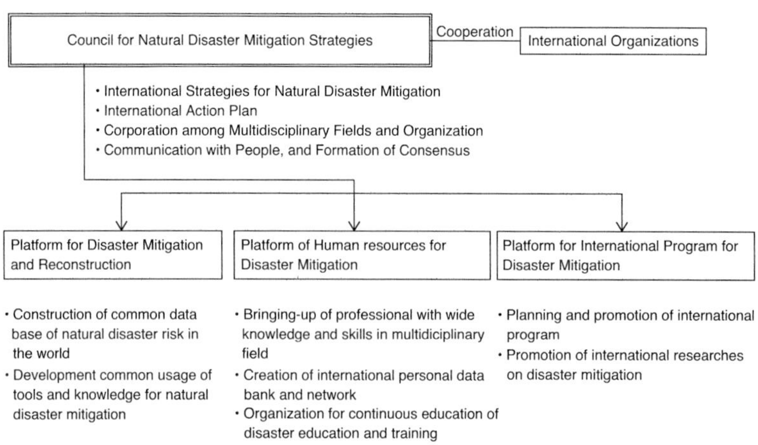

FIGURE 3.2.1
Proposal of an organization for international natural disaster mitigation strategies (Science Council of Japan).

3.2.2 Governmental Policy for Development of National Land Resilience in Japan

In December 2013, the Basic Law for National Land Resilience was enacted by the Diet of Japan, and the so-called "national land resilience" against natural disasters has been promoted in various fields under the basic policies of (i) protection of human life, (ii) maintenance of important functions of the nation and society, (iii) minimization of damage to people's property and public facilities, and (iv) prompt recovery and reconstruction.

In addition, the "Three-Year Emergency Measures for Disaster Prevention, Mitigation, and National Land Resilience" have been promoted since 2018, and the "Five-Year Acceleration Measures" have been promoted since 2021. The total amount of government expenditure for national land resilience by 2022 has reached about 50 trillion yen. As a result of these measures taken by the Japanese government, local governments, and private businesses, the risk of natural disasters in Japan has been steadily decreasing. However, the complacency and overconfidence was keenly felt by the 1995 Kobe earthquake, the 2011 East Japan earthquake, and the 2024 Noto Peninsula earthquake. National land resilience has been carried out mainly through projects that have been under the jurisdiction of each ministry and agency. Although the Cabinet Office plays a role in coordinating the overall situation, coordination among ministries and agencies is not sufficient, and it is necessary to check whether the country as a whole is effectively making progress in national resilience and whether there are any gaps in resilience measures.

On the contrary, in addition to earthquake and tsunami disasters, weather disasters such as storms, heavy rains, and floods, which are thought to be caused by global warming, are increasing around the world. According to statistics during the 70 years since 1950, there have been 121 natural disasters worldwide, each of which killed more than 1,000 people, resulting in 5.12 million casualties. In Asia, 89 natural disasters have killed 3.95 million people, and the majority of the world's natural disasters occur in Asia. The concentration of natural disasters in Asia can be attributed to excessive population concentration in urban areas, fragile living spaces, and delays in the development of disaster prevention infrastructures.

The "protection of human life," which is the first policy of the Basic Law for National Resilience in Japan, is the most important issue in Asia. Every time an earthquake occurs in developing countries such as Asia, we have repeatedly seen the sight of many residents dying because they were buried in the rubble of collapsed houses. As an expert in disaster prevention, I was gripped by a sense of regret that we had repeated the same faults. Earthquake-resistance enhancement of dwellings with adobe bricks and other materials has hardly progressed due to cost and technical problems. The downward spiral continues, in which poverty increases the degree of disaster, which in turn increases the degree of poverty.

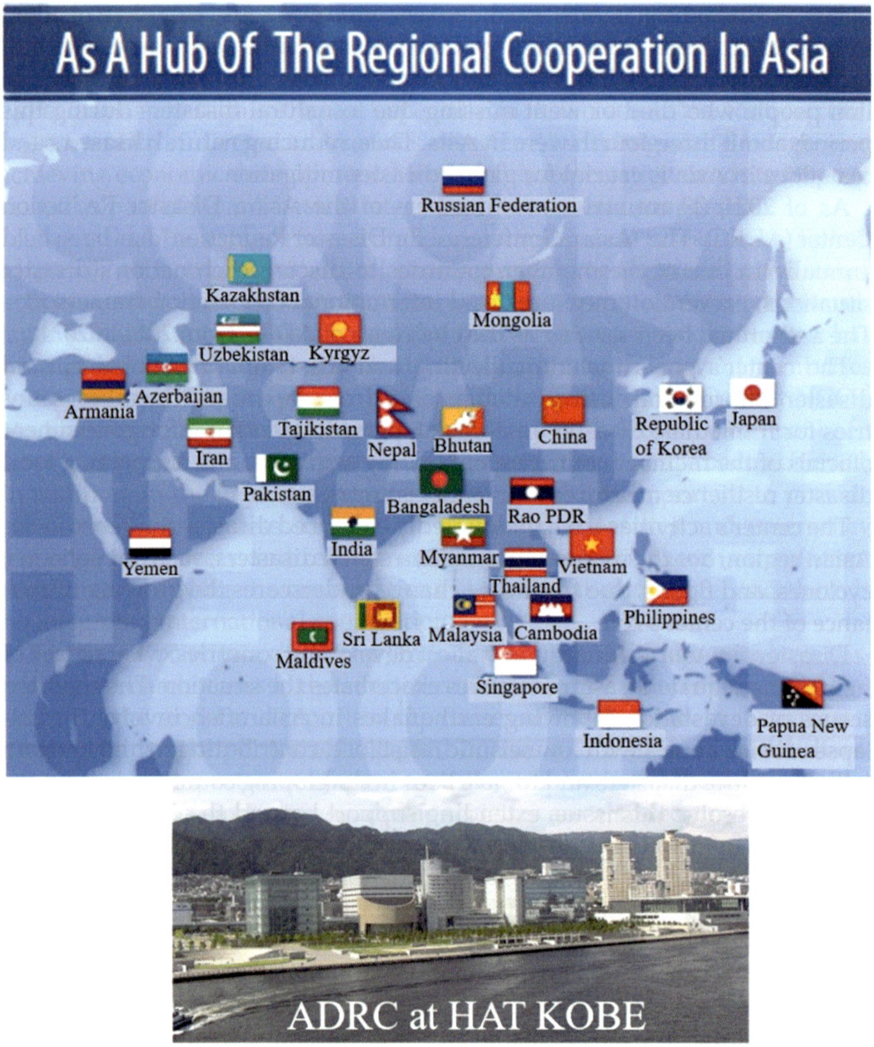

FIGURE 3.2.2
The Asian Disaster Reduction Center.

During a visit to the United Nations office opened in Banda Aceh, I explained that the purpose of the Japanese survey team was to support the formulation of reconstruction plans for Sumatra Island, particularly the restoration plan for the west coast road facing the Indian Ocean, which had suffered extensive damage. The UN officials immediately agreed to support our survey by providing a helicopter free of charge (Figure 3.2.3).

I learned from UN staff that a survey team from Engineers without Borders, USA, had already arrived at Banda Aceh to begin damage survey and recovery support. Engineers without Borders, USA, is an organization of civil and

FIGURE 3.2.3
Survey on the west coast of the Northern Sumatra.

architectural engineers that not only supports disaster recovery and reconstruction worldwide but also transfers civil and architectural technologies during non-disaster times. Similar organizations are promoting activities in Europe and other regions.

Japan has dispatched many survey teams worldwide following natural disasters, documenting damage and sharing findings primarily in Japanese, limiting the audience to specific institutions and organizations. As an engineer from a country that prides itself on being advanced in natural disaster prevention, the author felt a strong concern that merely documenting and analyzing damage was not enough and that Japanese technologies could be utilized to prevent natural disasters and to support recovery and reconstruction for affected areas and their residents.

Inspired by learning about Engineers without Borders during our visit to the UN office, I considered forming a similar organization in Japan. Upon returning to Japan, the author discussed the idea of establishing "Engineers without Borders, Japan" with executives from construction companies and consulting firms, receiving strong support. The boards of the Japan Society of Civil Engineers and the Architectural Institute of Japan also decided to support the initiative. Thus, Engineers without Borders, Japan (EWBJ), was launched on November 1, 2014.

EWBJ sent engineers to Ofunato and Rikuzentakata cities for reconstruction support following the 2011 East Japan earthquake and has conducted disaster prevention lectures and publications to reduce earthquake and tsunami disasters.

EWBJ is also supporting university students for their disaster educational activities abroad (Figure 3.2.4).

FIGURE 3.2.4
Lecture on natural disaster prevention at high schools in Sumatra.

Most members of the JSCE survey team conducted investigations such as tracing the tsunami runup height and estimating the tsunami forces from the state of building's damage. The survey results were reported by the Japan Society of Civil Engineers and the Architectural Institute of Japan.

Many children lost their lives and their families in the Sumatra tsunami. Most survivors of children had been saved and cared for in schools and hospitals in the suburbs. Some members of the Japanese team visited several elementary, junior high schools and an orphanage house. At an orphanage, the survey team members shared Japan's experiences of tsunamis and methods to protect lives. After the presentation by members of the survey team, a girl stood up and asked, why you had yet to share the knowledge and experience of tsunamis and evacuation methods with us before the tsunami struck. All members of the Japanese team lost their words to respond. I later learned from the orphanage staff that the girl had lost her parents and siblings in the tsunami.

Back in Japan, I shared the story of the girl from Banda Aceh with my students at the end of a lecture, which received a significant response. Waseda University students initiated a group to support disaster education domestically and internationally. Every summer, they visit Indonesia and the Philippines to conduct disaster education activities for elementary school children and junior high school students, funded by their part-time jobs during the summer holidays. These student-led disaster education activities also take place domestically, including creating education materials based on the evacuation experiences of elementary school children during the 2011 East Japan earthquake. Picture books have been produced in Japanese for domestic disaster education and in English and Indonesian for international activities. Engineers without Borders supports these student activities in terms of safety and finances. A student organization for disaster education was also formed at Kyoto University in 2005, and students from both Kyoto

FIGURE 3.2.5
Picture book for disaster education by student members of Engineers without Borders, Japan.

and Waseda universities closely collaborate on educational materials and the tasks for disaster education on-site.

About a year after the 2004 disaster, I visited Banda Aceh again. While walking along the tsunami-stricken coast, I heard the lively voices of children playing in the water. I spontaneously greeted them with "Boh aba," which means "Hello" in the Acehnese dialect. The children responded with "Boh aba." I thought that recovery from the disaster, especially the care for the children's hearts, was making slight progress (Figure 3.2.5).

3.3 International Corporations for Disaster Mitigation and Reconstruction

3.3.1 1990 Ruzon Island Earthquake, Philippines

On July 16, 1990, a magnitude 7.8 earthquake struck Luzon Island in the Philippines, causing widespread liquefaction in the central region of the island, including the city of Dagupan. This led to the collapse and tilting of buildings, the falling of bridge girders, and the rupturing of embankments, along with extensive damage to lifelines such as water and sewage systems.

Dagupan City, located about 200 km northwest of Manila and facing the Lingayen Gulf to the north, is a city with a population of approximately 130,000 (see Figure 1.5.6). The city features sandbars and dunes along its coastline, with many fishponds scattered in the marshlands between the dunes. The city center was constructed by reclaiming these marshlands.

As an aside, during JSCE's support activity for the recovery of buildings that had tilted and sunk due to liquefaction in Dagupan City, some influential locals invited us to dine at a restaurant. We were served sashimi made from a fish raised in the local fishponds. I don't remember the name of the fish, but it resembled a carp. Having suffered severe diarrhea caused by drinking unboiled water during a previous visit to Luzon Island right after the earthquake, I was hesitant to eat, but one of the influential guests, a medical doctor, assured us it was "absolutely safe," so I tried it. The fish was a freshwater species and had a refreshing taste. Fortunately, I did not experience any health issues afterward.

In Dagupan City, liquefaction caused several damages, such as sinking and tilting of buildings, fall of road bridges, rupture of water pipes, and the horizontal movement of the liquefied ground (Figures 3.3.1 and 3.3.2).

A five-story reinforced concrete building of a private university in the city, constructed along the Pantal River in the city, had sunk and tilted due

FIGURE 3.3.1
Tilted and subsided by liquefaction (Dagupan City, 1990 Philippines earthquake).

FIGURE 3.3.2
Fall of a bridge by liquefaction-induced ground displacement.

to liquefaction (Figure 3.3.3). The building had sunk by as much as 150 cm, and the maximum tilt angle was about 3°. The university's management sought technical cooperation for the building's restoration. Supporters from Engineers without Borders, Japan shipped 30 jacks from Japan and dispatched a specialist for the restoration of the building.

The recovery method is illustrated in Figure 3.3.4. The procedure of construction is as follows: (i) cut and detach all the concrete columns on the first

(a)

(b)

FIGURE 3.3.3
Inclined and sunk building by soil liquefaction (1990 Philippines earthquake): (a) a five-story building tilted and sank due to liquefaction and (b) the first floor of the titlted and sank building.

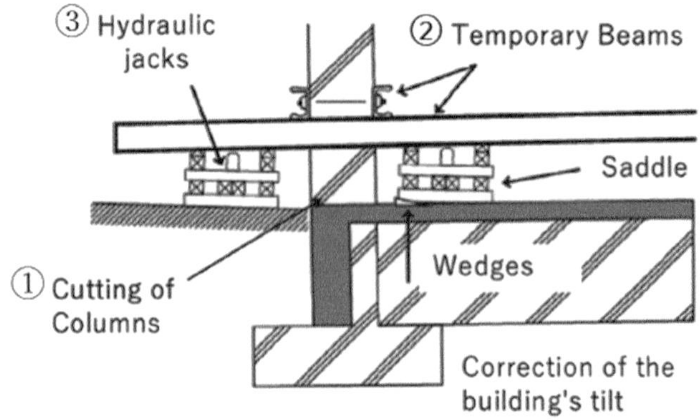

FIGURE 3.3.4
Restoration work for inclined building.

floor, (ii) install temporary beams between the detached columns to receive the jack reaction force, (iii) place 30 jacks between the foundation and the temporary beams to raise the sunken side, and (iv) cast new reinforcement concrete pillars between the detached columns.

Local construction engineers were invited to participate in the recovery work. After completing the recovery work, all the jacks were donated to Dagupan City, hoping to use them in the city's reconstruction by transferring the recovery technology. However, I have yet to learn how the donated jacks have been utilized since then.

3.3.2 2008 Wenchuan Earthquake, China

On May 12, 2008, the central section of the Longmen Shan Fault, located at the boundary between the Sichuan Basin and the Tibetan Plateau in China, ruptured over a length of 250 km, triggering an earthquake with a magnitude 8.0. According to the Chinese government, this disaster resulted in approximately 70,000 deaths, about 18,000 missing persons, and the collapse of over 5.3 million houses and buildings. The earthquake's epicenter was in a mountainous region, leading to landslides, the formation of lakes due to debri-blocked rivers, the destruction of buildings and tunnels intersecting with the fault, and the collapse of bridges. In urban areas like Chengdu, the capital of Sichuan Province, a huge number of concrete block houses and buildings collapsed, causing a number of fatalities (Figure 3.3.5).

In the field of earthquake disaster prevention, Japan and China have maintained a cooperative relationship. Waseda University has received many Chinese students from China, who have gone on to hold key positions in China's central and local governments. One such individual is Dr. K, vice president of Southwest Jiaotong University in Chengdu.

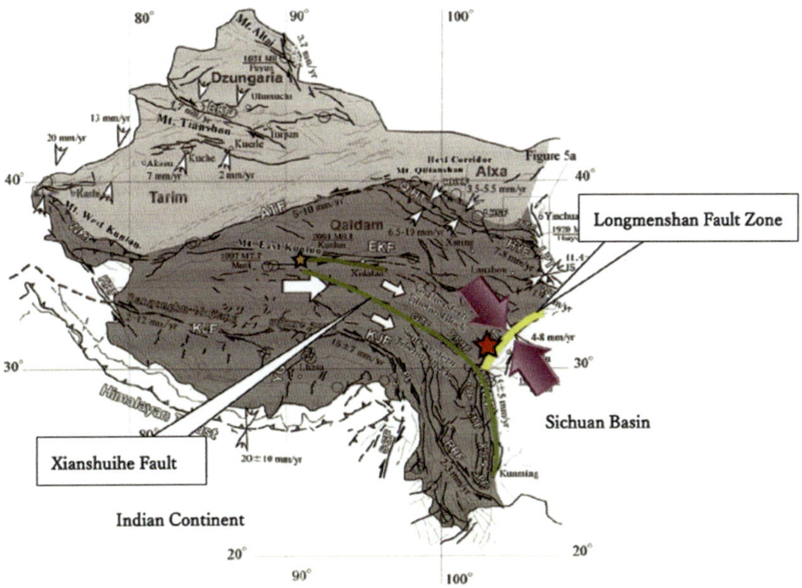

FIGURE 3.3.5
Plate technics of Sichuan Basin and Tibetan plateau.

Immediately following the Wenchuan earthquake, Dr. K called for cooperation from Japanese researchers to organize a joint team to survey the damage and to make reconstruction plans. In Japan, a team for damage assessment and recovery was organized jointly by Japan Society of Civil Engineers (JSCE) and Architectural Institute of Japan (AIJ), with Engineers Without Borders Japan (EWBJ).

The team's initial focus was on the construction of partially damaged buildings and bridges, excluding completely collapsed structures. For structures that could be almost restored to their original form through recovery efforts, restoration methods were discussed with Chinese engineers and reported to the managers and owners of the structures (Figure 3.3.6).

After the 1995 Kobe earthquake, partially damaged buildings and structures were restored to their original state and reused. For the Wenshuan earthquake, plans were proposed to select several damaged structures for recovery, one of which was a six-story concrete building (Figure 3.3.7).

The building was damaged at column-beam joints on the first floor which could lead to total collapse if strong aftershocks occurred. The reinforcement method involved wrapping the columns with steel plates and filling the space between the steel plates and the damaged concrete columns with new concrete as shown in Figure 3.3.8. This method has been used after the Kobe earthquake for the recovery of concrete buildings, and seismic reinforcement of Shinkansen viaduct columns, with its effectiveness confirmed through full-scale experiments. This marked the first time a recovery method developed in Japan was applied to structures damaged overseas after Kobe earthquake.

FIGURE 3.3.6
Japan-China technical meeting for restoration of damaged infrastructures (2008 Wenchuan earthquake, China).

FIGURE 3.3.7
Joints between concrete columns and beams were ruptured (2008 Wenchuan earthquake, China).

In May 2008, special joint lectures by experts in geotechnical engineering, seismology, and earthquake engineering from both Japan and China were held at Southwest Jiaotong University. The goal was to transfer Japanese seismic engineering techniques in civil engineering and architecture to young engineers and students in China.

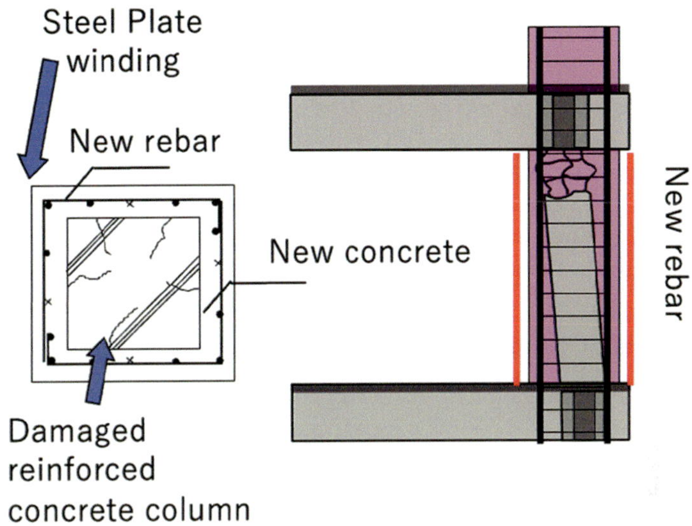

FIGURE 3.3.8
Restoration of damaged columns and its joints with beams of building.

The main themes of the special lectures were given as follows:

1. Faults, earthquake predictions, and ground motions
2. Seismic design and reinforcement of buildings
3. Seismic reinforcement techniques for bridges, embankments, and other road structures, dams, and tunnels
4. Liquefaction countermeasures
5. Slope stability design and countermeasures

The damaged concrete piers were winded by steel plates, and concrete was cast between plates and existing piers.

Following the special lectures, with support from the Chinese government, the "Wenchuan Province Key Laboratory of Earthquake Engineering" was established on the campus of Southwest Jiaotong University, equipped with various experimental facilities, including some large and high-performance ones.

Financial support from the Japan Bank for International Cooperation (JBIC) was obtained to conduct the special lectures. After learning about our activities in Chengdu through media reports, JBIC offered to support the lecture series, covering travel and accommodation expenses for the lecturers. The special lectures in Chengdu were initially planned to continue for several years, targeting young researchers and students from universities in southwestern China.

However, JBIC's support was discontinued after one year. According to JBIC officials, the reason for the discontinuation was that there were no plans for a support program for the Wenchuan earthquake in the fiscal year 2008, so it was decided to support the activities of the JSCE and EWBJ for only one year. From the next fiscal year, JBIC will implement its own programs and, therefore, will not be able to support the special lectures. The special lectures in the 2009 fiscal year had to be significantly scaled down. It was regrettable, especially since collaborative relationships were established between Chinese universities and Japanese researchers.

3.3.3 Proposal of Tsunami Warning System in Sumatra, Indonesia

After the 2004 Sumatra offshore earthquake, there were prolonged aftershock activities, including some exceeding moment magnitude 8. The people of Banda Aceh of Sumatra Island, who had suffered severe damage, retained a strong fear of tsunamis. When the shaking from aftershocks began, they would rush outdoors to flee from possible tsunamis, causing great chaos. Many residents attempted to evacuate to higher ground by car, leading to significant traffic congestion. This chaos resulted in additional fatalities.

Disaster prevention education for elementary and junior high school students conducted by Engineers Without Borders Japan (EWBJ) became a catalyst for people to take quicker evacuation actions when aftershocks occurred. However, due to funding issues, there needed to be more progress in establishing a tsunami warning system and constructing evacuation routes and shelters.

The EWBJ, the survey teams from the Japan Society of Civil Engineers and the Architectural Institute of Japan proposed the development of a tsunami warning system to the Indonesian government and local authorities. The initial response had not been active due to the priority given to the vast works of recovery and reconstruction projects.

The situation changed after human lives were lost due to the aftershocks. The need for a tsunami warning system was recognized anew, and requests for proposals tailored to the local Islamic society were conveyed to Engineers without Borders by officials from Banda Aceh.

Considering the current state where no seabed seismograph was installed off the west coast of Sumatra, and the number of seismic observation points along the west coast was limited, as well as the lack of a database on coastal topography, which significantly affects the height and arrival time of tsunamis, the challenge was how to send warnings to residents and encourage early evacuation. In Japan, warnings from the Japan Meteorological Agency are conveyed to residents through TV, radio, smartphones, and local disaster prevention wireless systems. Constructing and operating a similar system in Sumatra in the short term seemed difficult.

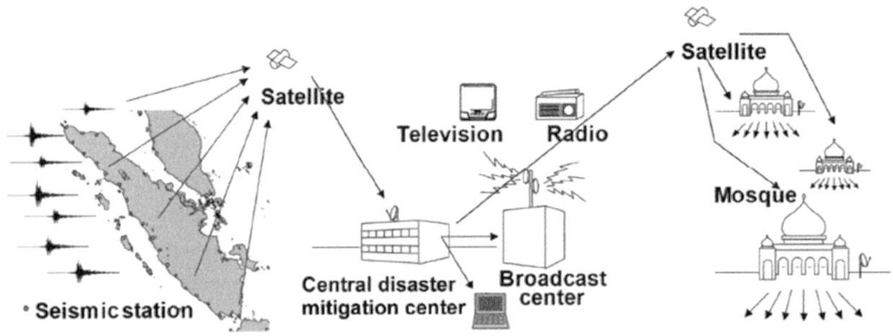

FIGURE 3.3.9
Tsunami evaluation and warning transmission system in Sumatra.

Taking these realities into account, a tsunami warning system utilizing satellites and mosques, which are built in each settlement in Islamic society, was proposed, as illustrated in Figure 3.3.9. Additional seismometers would be installed along the west coast of Sumatra to catch earthquake motions promptly. Due to the high cost of installing seismometers, realizing this with support from EWBJ alone would be challenging. Efforts would be made to lobby the Japanese government and international organizations like JICA for financial support. The maintenance of the seismometers will be managed by local authorities along the coastline.

When an earthquake occurs in the Indian Ocean, multiple seismometers installed along the coastline will detect the seismic motions. These data would be transmitted to the Tsunami Disaster Mitigation Center (provisional name) via satellite, which would then automatically estimate the epicenter and the magnitude of the earthquake. The center, located in major urban areas like Jakarta, would estimate the tsunami's arrival time and height along the coastline based on the earthquake magnitude and the tsunami sources, then send the information to broadcasting stations and, again through satellites, to local mosques.

In Islamic society, mosques are built in each settlement. Tsunami warnings would be conveyed to residents through speakers on the mosque's minarets. The mosque broadcasts verses from the Quran 5 times a day, and residents pray in accordance with these broadcasts. Conveying tsunami warnings through mosques would be a system that leverages the lifestyle of Islamic society, where mosques are located in the cente of the community. Many mosques withstood the tsunami caused by the Sumatra earthquake, saving the lives of residents who took refuge in them (see Figure 1.3.5). Using mosques as evacuation centers and stockpiling food and daily necessities could significantly aid in the relief of evacuees after disasters.

3.3.4 Reinforcement Plan for the Tomb of Amenhotep III in Egypt[21]

When I was working at the Faculty of Science and Engineering at Waseda University, Professor Yoshimura Sakuji, renowned for his work in Egyptology, visited my lab. Originally from Waseda's Faculty of Letters, Arts, and Sciences, he was also part of the civil and architectural engineering group at the Research Institute for Science and Engineering, making him a colleague of mine.

Professor Yoshimura was working on the restoration of wall paintings in the tomb of Pharaoh Amenhotep III located in the Western Valley of the Kings in Luxor, Egypt, which he was conducting at the behest of UNESCO. Volunteers from Waseda's architectural department were participating in this restoration project. Amenhotep III's tomb, carved from limestone, was discovered in 1799 during Napoleon's expedition to Egypt. The underground tomb's columns and walls had been cracked and delaminated, with some columns completely collapsed, possibly due to past floods or earthquakes. The Egyptian Antiquities Authority was concerned that if left unaddressed, the tomb and its valuable wall paintings might be lost. They requested the Japanese research group to propose a reinforcement method.

At the time, I was conducting a survey on abandoned lignite mines in Mitake Town, Gifu Prefecture, Japan (Collapse of underground cavern), and had been involved in developing seismic reinforcement methods as well as conducting reinforcement works for civil engineering structures damaged by the 1995 Kobe earthquake. Intrigued by the project, I accepted the survey request. I enlisted the help of Professor Omen Aidan from Tokai University's Faculty of Marine Science and Professor Tano from Nihon University's Faculty of Science and Engineering, both experts in rock mechanics and familiar with the safety of rock caverns, having collaborated with me on the of lignite cavern survey.

Amenhotep III, a Pharaoh of the 18th Dynasty of Ancient Egypt and Tutankhamun's grandfather, ruled for nearly 40 years during a time of prosperity. Preparing hurriedly, we flew from Kansai Airport to Luxor via Istanbul and Cairo. Guided by Assistant Professor Kondo from the Faculty of Letters of Waseda University, we headed to the Valley of the Kings. While many tombs in the Eastern Valley are open to the public, the Western Valley, where Amenhotep III's tomb is located, is not, as it's reserved for academic research and restoration. A concrete building named "Waseda House" at the entrance of the Western Valley has served as the base for Waseda's Egyptological research since 1989.

After resting for one night at Waseda House, we entered the Western Valley early the next morning. Napoleon III's expedition discovered the tomb's entrance halfway up a slope covered in sediment. Like Tutankhamun's tomb, discovered accidentally by Dr. Howard Carter, finding the entrance to underground royal tombs is the starting point of archaeological surveys in Egypt. New tombs are speculated based on the alignment of known entrances and topography. The author tried various methods to locate abandoned lignite mine cavities from the ground surface in Mitake Town, Japan. Seismic and radar surveys proved somewhat effective for limited depths, suggesting their potential to discover new royal tombs when combined with topographical studies.

Inside Amenhotep III's tomb, we examined the condition of the columns and walls adorned with valuable wall paintings. Specialists from Italy were restoring the paintings, a labor-intensive and meticulous task. We thoroughly investigated the cracks and delamination in the columns and walls, applying Japan's experience in repairing concrete structures damaged by past earthquakes, such as the 1995 Kobe earthquake in Japan.

Several reinforcement methods were proposed, including encasing columns and beams with steel plates as shown in Figure 3.3.8, constructing new reinforced concrete columns around the existing ones, injecting adhesive into cracks, and fastening the damaged columns with steel bolts. However, the Egyptian Antiquities Authority rejected the use of artificial materials like steel and concrete, which are dissimilar to natural rock. Finally, carbon fiber belts to fasten the crack of the columns and beams were accepted (Figures 3.3.10 and 3.3.11).

However, there has been no response from UNESCO. It was later revealed that the majority of the funding was expected to come from the Japanese government. Over 200 years had passed since the tomb's discovery without further collapses, and there was no immediate threat of collapse; UNESCO seemed to prefer taking more time to consider the restoration method.

While exploring the valley, I found seashell fossils on a hilltop in the valley, indicating that the area was once undersea, a fact confirmed by geologists. A similar thing was experienced during the damage survey of the 1999 Wenchuan earthquake in China, where the collision of the Indian subcontinent with Eurasia led to the uplift that formed the Himalayas and the Tibetan Plateau, far exceeding the timescale of engineering.

FIGURE 3.3.10
Kings' Valley in Luxor, Egypt.

FIGURE 3.3.11
Damaged column of the tomb of Amenhotep III.

4

Construction Methods for Resilient Infrastructures against Earthquakes and Tsunamis

4.1 Soil Improvement

4.1.1 Sand Compaction Pile

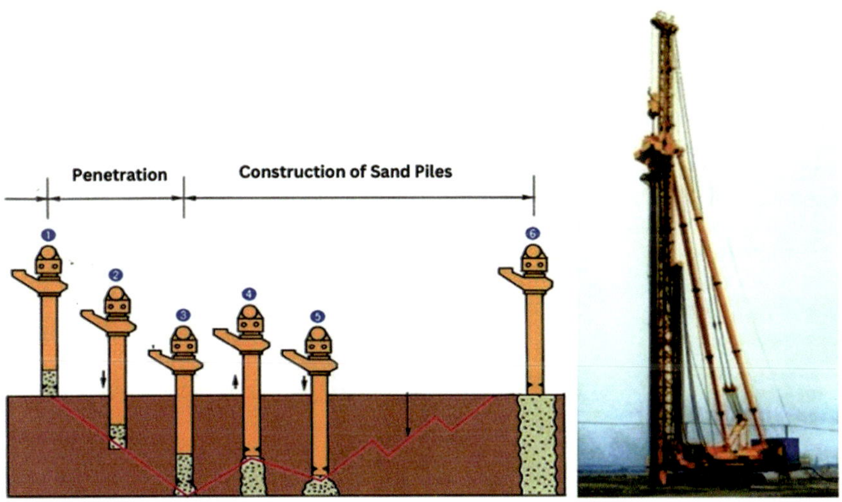

FIGURE 4.1.1
A vibratory hammer is used to penetrate a casing into the ground, and the casing is pulled out at a predetermined depth to expel sand and other materials, which are then driven back to create a well-compacted sand pile. For loose sandy soil, the entire ground is compacted by increasing the density, and for cohesive soil, the ground is strengthened by forming a composite of cohesive soil and sand piles. It is applied both on land and at seabed. (Data Provider's Affiliation: Fudo Tera Co.)

DOI: 10.1201/9781003728788-4

4.1.2 Grouting

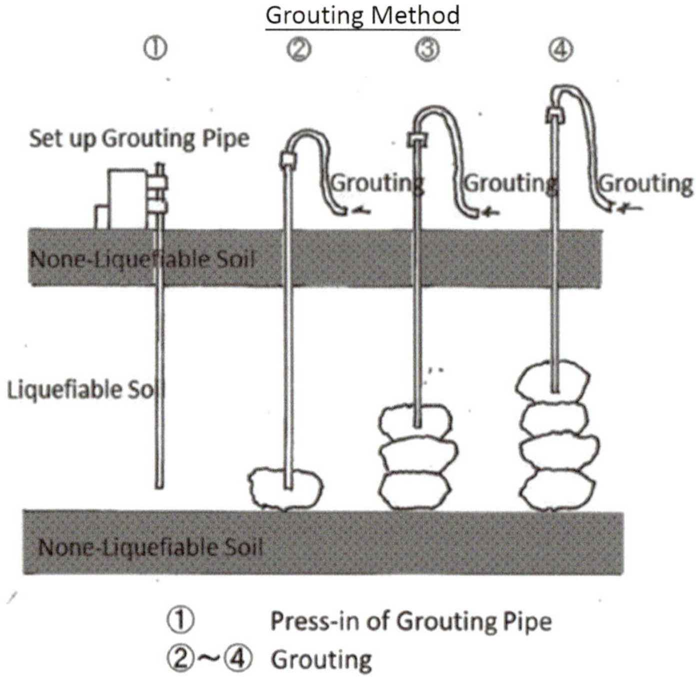

FIGURE 4.1.2
Solidifying materials, such as cement milk, are injected into the ground through pipes to construct solid ground columns to prevent liquefaction and its induced ground displacement. (Data Provider's Affiliation: Tobishima Construction. Reference Links: https://www.tobi-tech.com/tech/carbonstock.html.)

4.1.3 Timber Piles

FIGURE 4.1.3
The logs do not deteriorate below the groundwater level. The log head is covered with soil to prevent bio-deterioration, thus ensuring durability. The logs are pressed into the ground to increase soil density and liquefaction resistance. At the same time, the logs create an underground storage of CO_2 and contribute to reduce global warming.

4.1.4 Deep Mixing

FIGURE 4.1.4
Improvement of up to diameter 2,500 mm is possible. High-capacity agitator heads can be used for hard ground. Forward and reverse rotation mechanism is effective for high-performance agitation and enables high-strength and homogeneous improvement. Smooth mud discharge and minimal impact on surrounding structures. (Data Provider's Affiliation: Raito Kogyo Co. Reference Links: https://www.raito.co.jp/.)

4.1.5 Gravel Drain

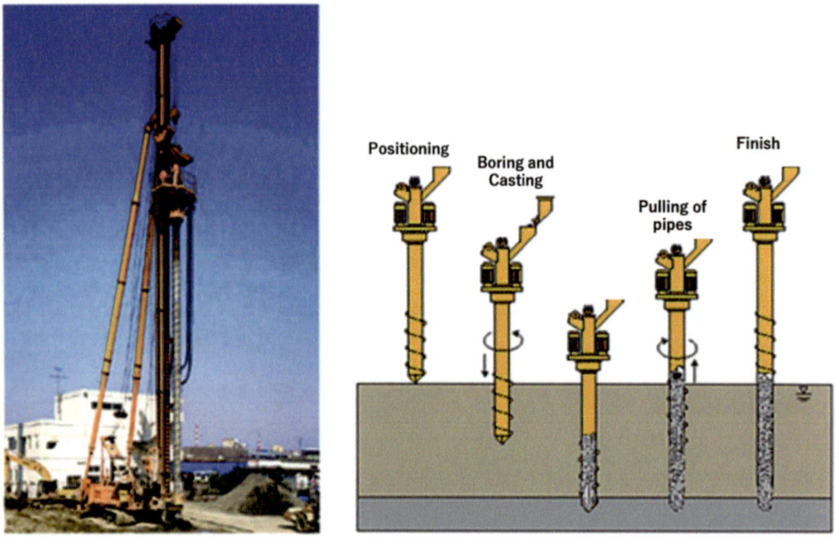

FIGURE 4.1.5
The gravel drain method improves the drainage of the ground and prevents liquefaction by suppressing excessive pore water pressure during earthquakes. The casing is rotated by an auger while penetrating the ground, and a gravel drain with high permeability is constructed in the ground.

4.1.6 Weight Dropping

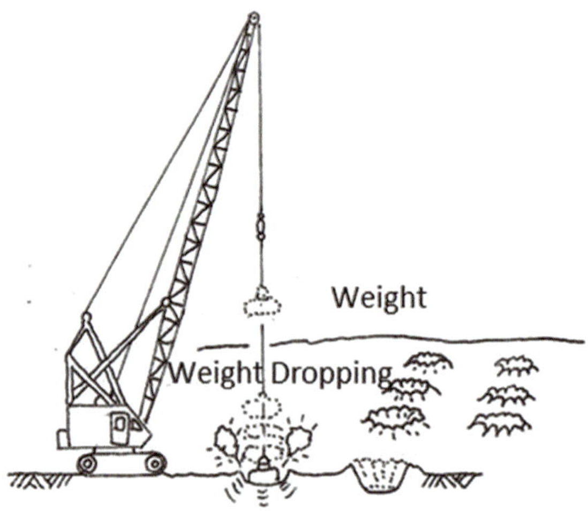

FIGURE 4.1.6
A weight is dropped from the air, and the impact force on the ground surface compacts the ground. The improved depth is limited to 2–3 m below the ground surface. The method is effective for the construction of lightweight structures.

4.2 Reinforcement of Foundations

4.2.1 Injection of Solidifying Materials

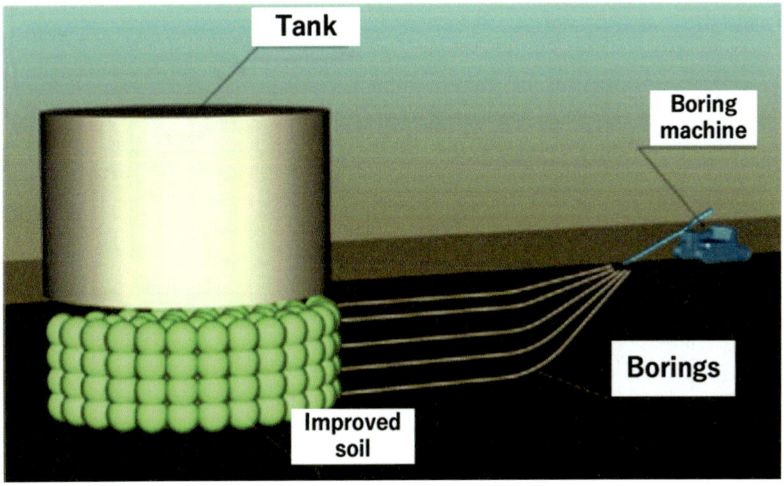

FIGURE 4.2.1
By infecting solidifying materials directly under the structures, the foundation ground is improved. The boring is controlled automatically. (Data Provider's Affiliation: Obayashi Co.)

4.2.2 Sheet Pile Wall

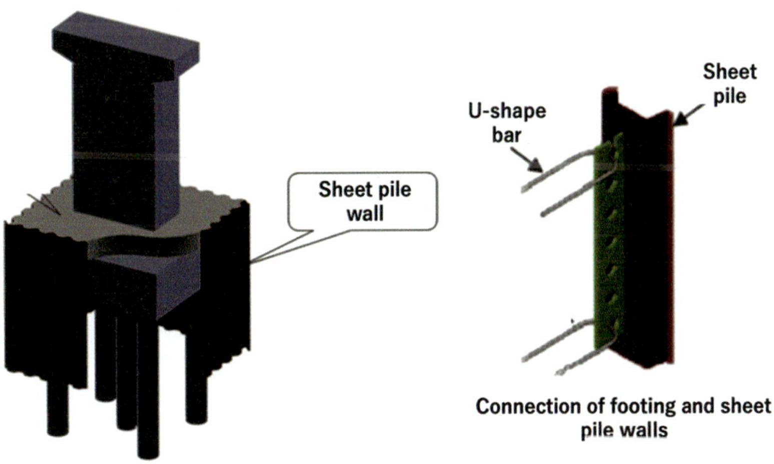

FIGURE 4.2.2
Steel sheet pile walls are constructed around existing footings. The foundation is reinforced by vertical and horizontal resistances of the steel sheet pile, and the ground horizontal displacement is prevented. (Data Provider's Affiliation: Design and construction manuals published by Railway Technical Research Institute, Obayashi Co. and Nippon Steel Co.)

4.2.3 Additional Piles

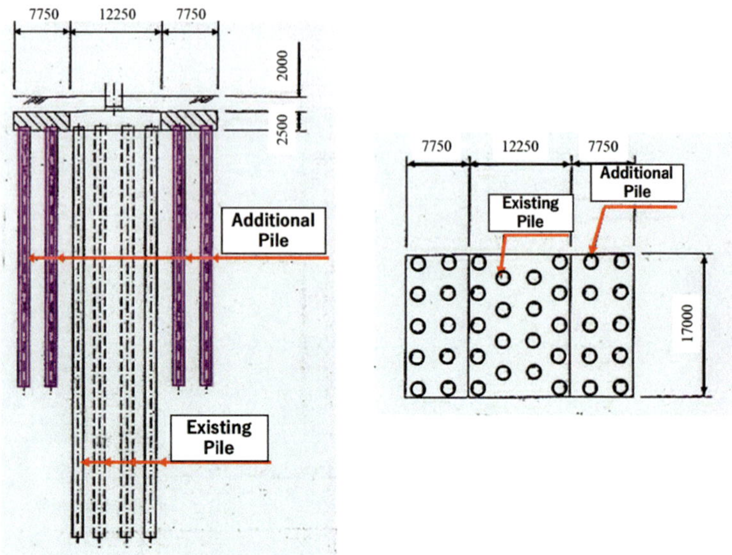

FIGURE 4.2.3
Additional piles are driven around the existing piles to increase the bearing capacity and rigidity of the foundation. Consideration must be given to securing construction space, new piles, and traffic obstructions.

4.2.4 Solidifications of Ground around the Existing Foundations

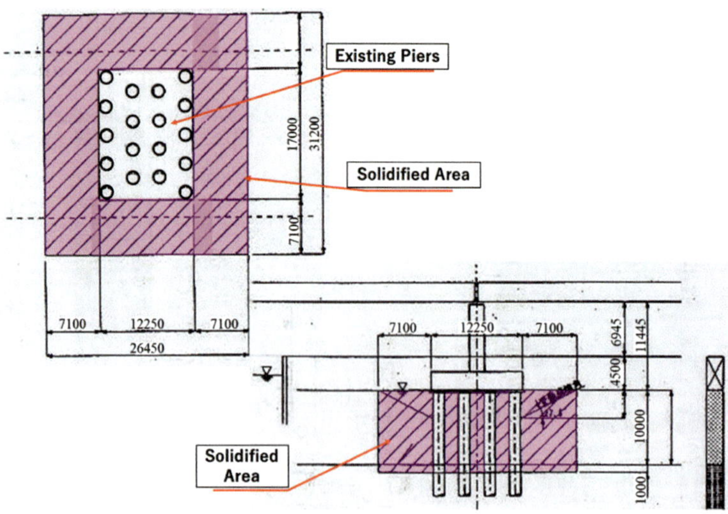

FIGURE 4.2.4
The ground inside and around the existing piles is strengthened by injecting solidifying materials.

4.3 Reinforcement of Sea Wall

4.3.1 Ground Anchor

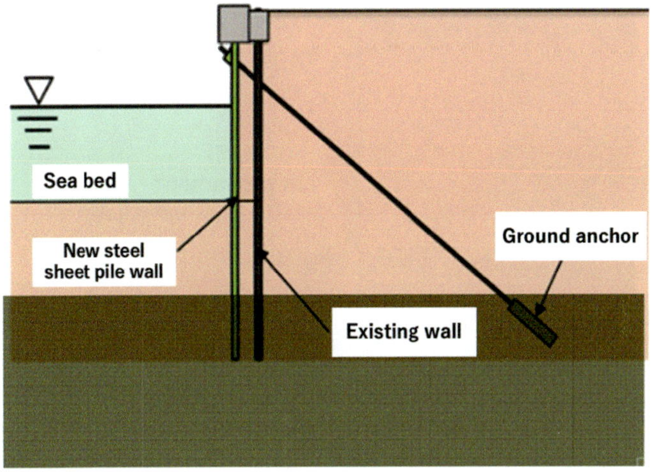

FIGURE 4.3.1
New steel sheet pile wall is constructed in front of the existing seawall, and the heads of the new walls are supported by ground anchors. (Data Provider's Affiliation: VSL Japan Reference Links: www.vsl-japan.co.jp.)

4.3.2 Continuous Steel Pile Wall

FIGURE 4.3.2
New steel pile wall is constructed in front of the existing sea wall to prevent the ground displacements behind the wall. (Data Provider's Affiliation: Giken Ltd. ("Gyropress Method" is a registered trademark of Giken Ltd and Nippon Steel Corp. in Japan.) Reference Links: www/giken.com/ja/.)

4.3.3 Soil Improvement of the Ground around the Existing Sea Wall

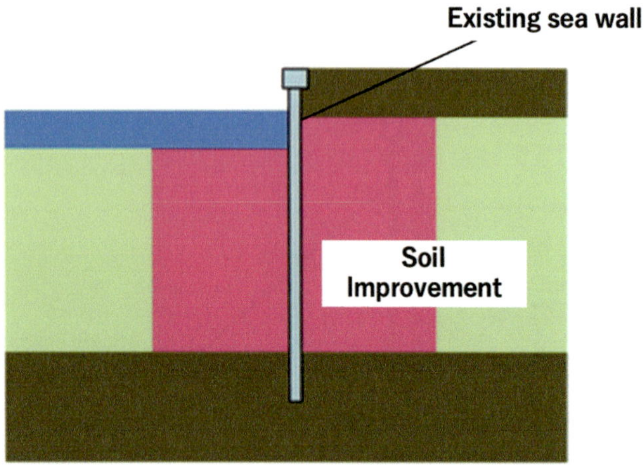

FIGURE 4.3.3
The ground in front of and behind the existing wall including the seabed is improved by injecting solidified materials.

4.4 Reinforcement of Concrete Colum

4.4.1 New Concrete Winding

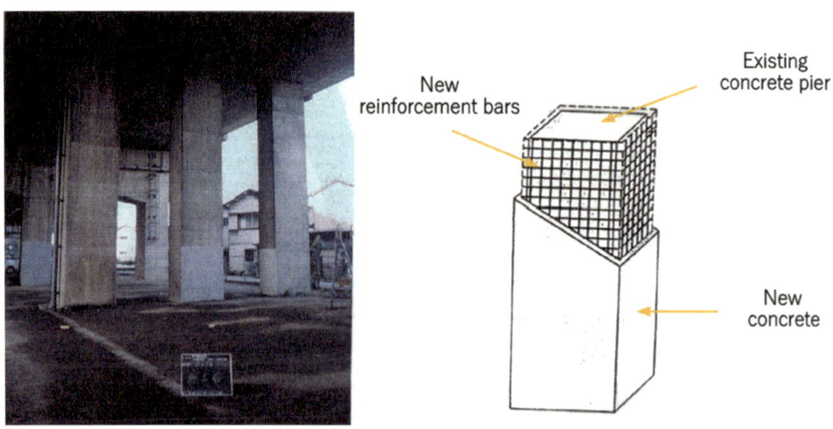

FIGURE 4.4.1
New concrete with reinforcement bars is cast around the existing concrete piers.

4.4.2 Steel Plate Winding

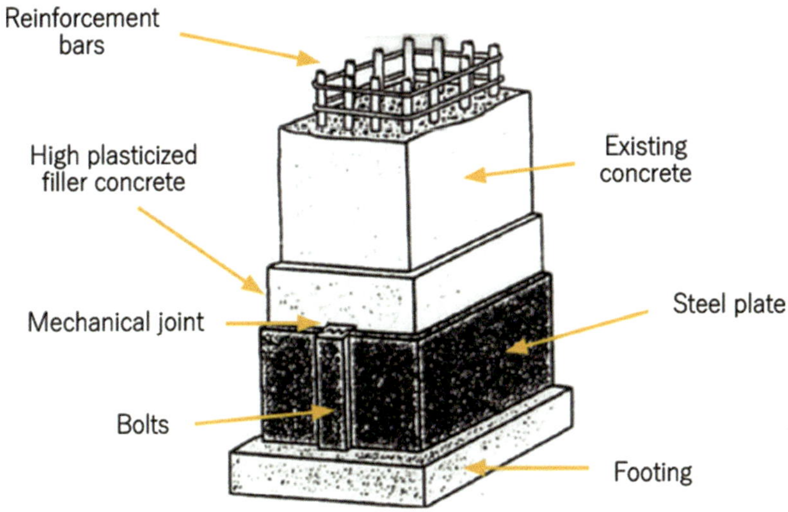

FIGURE 4.4.2
Steel plates are winded around the existing concrete piers, and the gap between the steel plate and the existing concrete is filled with high plasticized concrete.

4.4.3 Carbon Fiber Winding

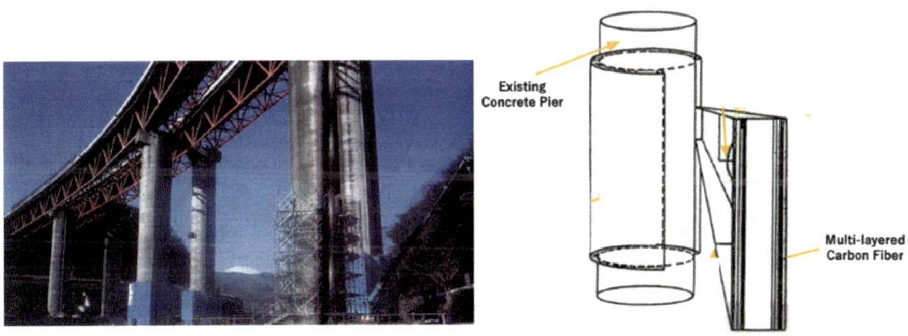

FIGURE 4.4.3
Carbon fibers are wrapped around existing concrete columns to prevent expansion due to shear failures of concrete and to make them more resilient structural members. Compared to concrete winding method, this method is more advantageous for high bridge piers.

4.4.4 Evaluation of Effectiveness of Reinforcement

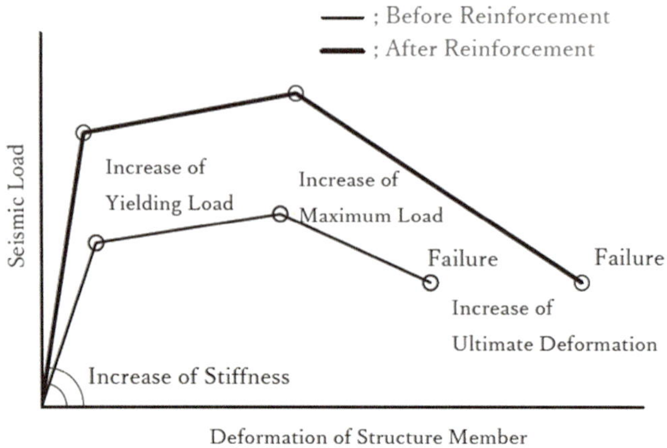

FIGURE 4.4.4
The effectiveness of reinforcement of structural members is evaluated as the increase of stiffness, yielding load, maximum load and ultimate load, and deformation at failures.

4.5 Reinforcement of Earth Dam and Embankment

4.5.1 Additional Embankment on Both Upstream and Downstream Slopes

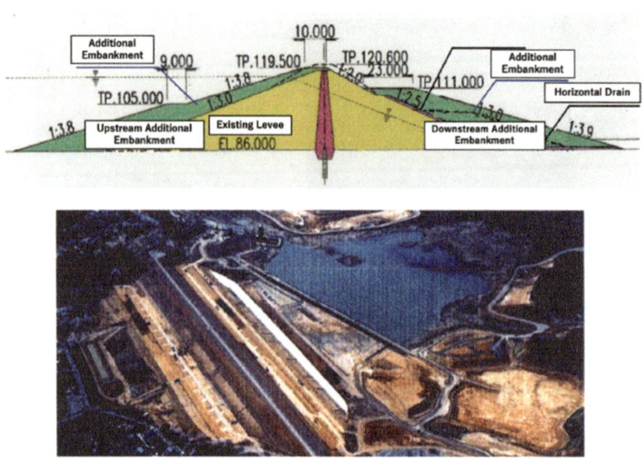

FIGURE 4.5.1
When the water level of the reservoir is low, the upstream side slope can be reinforced by additional embankment to increase the slope stability. Construction space is required for extension of the foot of the slopes.

4.5.2 Additional Embankment on Downstream Slopes

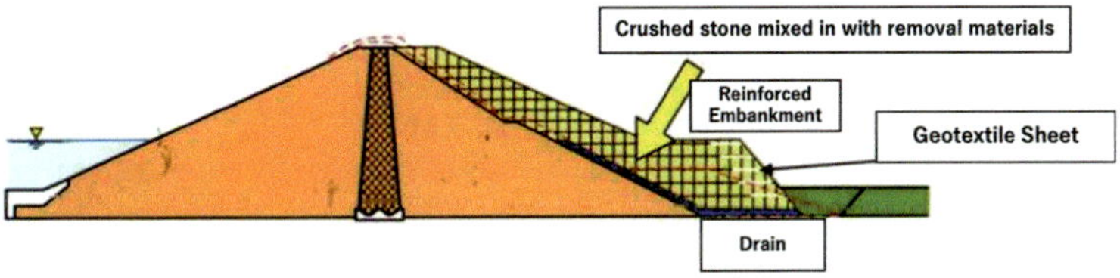

FIGURE 4.5.2
In the case when the upstream slope is difficult to reinforce, only the downstream slope is reinforced with geotextile sheet, which is effective to reduce the construction space.

4.5.3 Steel Sheet Pile Wall Driving at the Shoulder of the Embankment

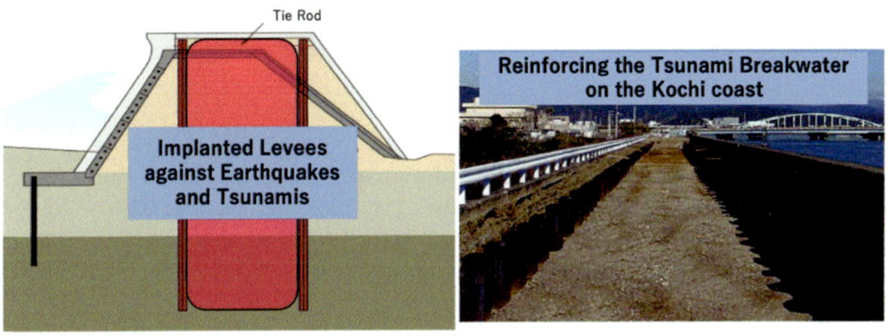

FIGURE 4.5.3
Sheet pile continuous wall is constructed along the shoulder of the embankment against earthquakes and tsunamis. The wall is expected to prevent the collapse of the wall and reduce and height and the force of tsunami.

4.5.4 Steel Pile Wall at the Foot of the Embankment

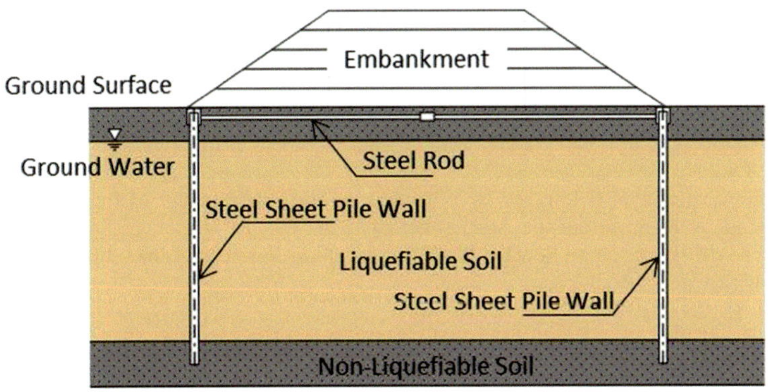

FIGURE 4.5.4
The wall is effective to prevent lateral movement of the ground under the embankment, resulting in the subsidence of the embankment.

4.6 Measures for Tanks and Residential Houses against Liquefaction

4.6.1 Groundwater Level Lowering (Tank Yard)

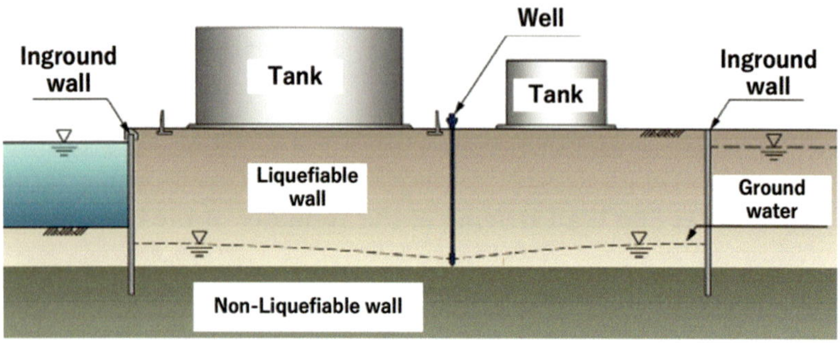

FIGURE 4.6.1
A tank yard ground is surrounded by a continuous impermeable sheet pile, and the groundwater level is lowered by pumping for the prevention of liquefaction. After the construction, it is usually required to watch the water level. (Data Provider's Affiliation: Giken Ltd. Reference Links: www.giken.com/ja/.)

4.6.2 Groundwater Level Lowering (Residential Area)

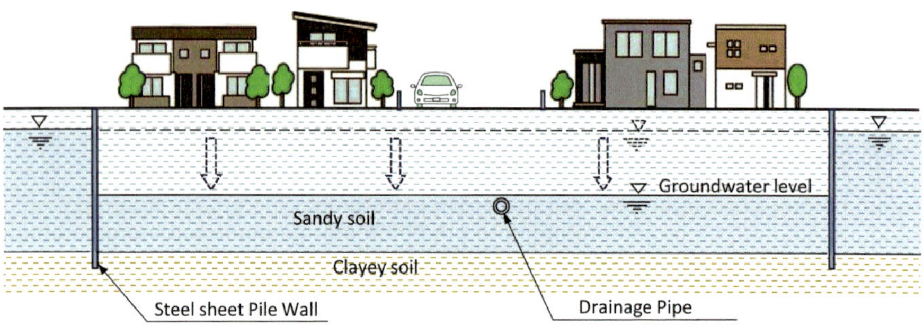

FIGURE 4.6.2
The groundwater level is lowered by pumping up the groundwater of the residential area surrounded by an impermeable steel sheet wall for the prevention of liquefaction. (Data Provider's Affiliation: Japan Sewage Works Association Reference Links: https://www.jswa.jp/en/jswa-en/.)

4.6.3 Skip-Driven Pile (See Section 2.2)

4.7 Measures for Inground Structures against Liquefaction

4.7.1 Improvement of Backfill Soil

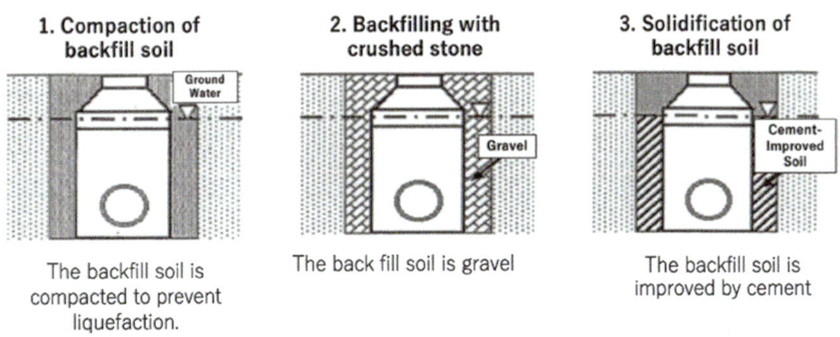

FIGURE 4.7.1

4.7.2 Prevention of Up-Lift of Sewer Manhole (Weight Increase, Drainpipe, Ground Anchor)

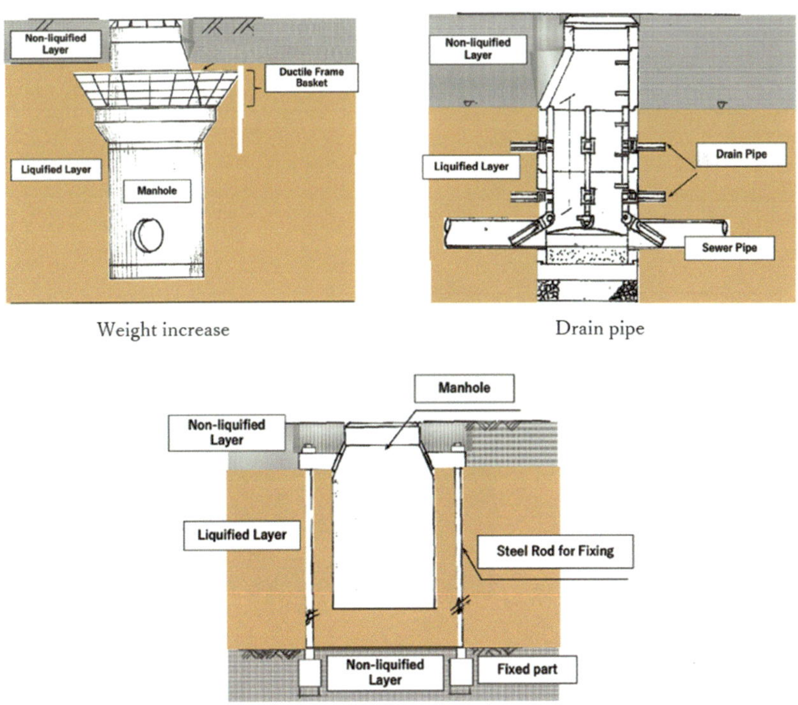

Weight increase

Drain pipe

Ground Anchor

FIGURE 4.7.2

4.8 Measures against Tsunami

4.8.1 Tsunami Protection Wall and Piles

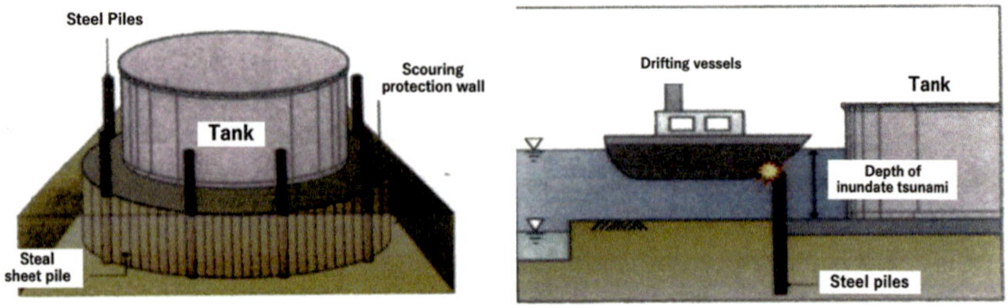

FIGURE 4.8.1
Steel sheet piles are driven around the tank foundation to prevent scouring of the ground due to tsunamis. To protect tanks against drifting objects, steel piles are driven at proper intervals. (Data Provider's Affiliation: Kajima Corporation Reference Links: https://www.kajima.co.jp/tech/c_soil_improvement/index.html#!body_02.)

4.8.2 Tsunami Measures for Nuclear Power Plants

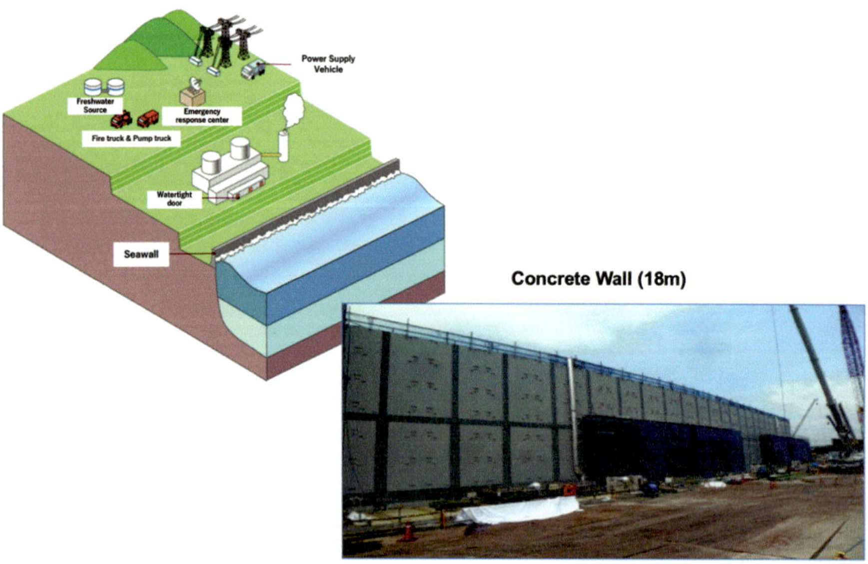

FIGURE 4.8.2
After the East Japan earthquake, tsunami protection walls have been constructed in all the nuclear plants in Japan. Furthermore, emergency equipment was moved to higher location.

4.8.3 Tsunami Evacuation Shelters

FIGURE 4.8.3
Learning from the lessons of the 2011 Great East Japan earthquake and the 1993 Hokkaido Southwest offshore earthquake, which resulted in a large loss of human lives, evacuation facilities have been constructed in coastal areas throughout Japan.

4.9 Measures against Surface Fault Movement

Measures for the Tokaido Shinkansen (the bullet train) were introduced in Section 2.3.1 (Figure 2.3.3), and a measure for earth dams for a concrete gravity dam was shown in Figure 2.3.4).

4.9.3 Measures for Water Supply System (East Bay Municipal Utility District, Oakland, USA)

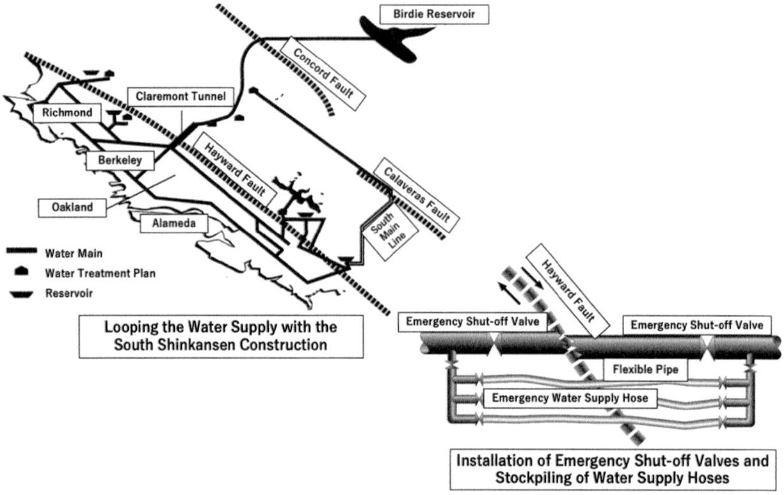

FIGURE 4.9.3
The EBMUD (East Bay Municipal Utility District) is an agency that supplies water to the area around Oakland in California, USA. Its water source is the Pardee Reservoir and the main water pipeline from the source to the supply area intersects with the active Hayward Fault. Anticipating damage to the water pipe due to the fault displacement, backup pipes have been stored nearby. An additional looped supply network (southern main line in the figure) was constructed.

4.10 Measures for Slope Stability

4.10.1 Slope Protection Pile

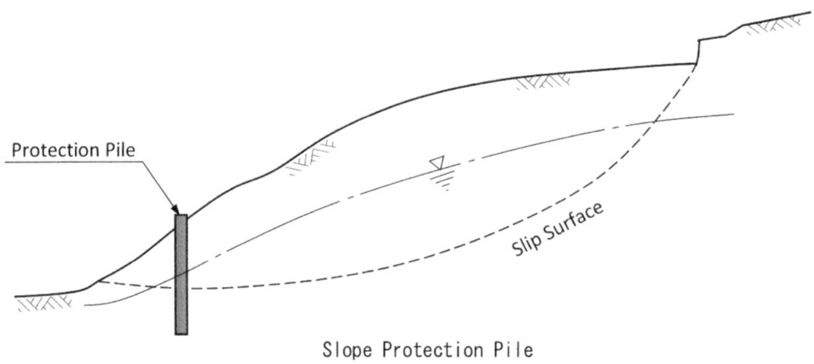

FIGURE 4.10.1
Steel piles are driven continuously at a proper interval at the foot of the slope. The deformation stability and the stress are examined against external earth pressure.

4.10.2 Ground Anchor

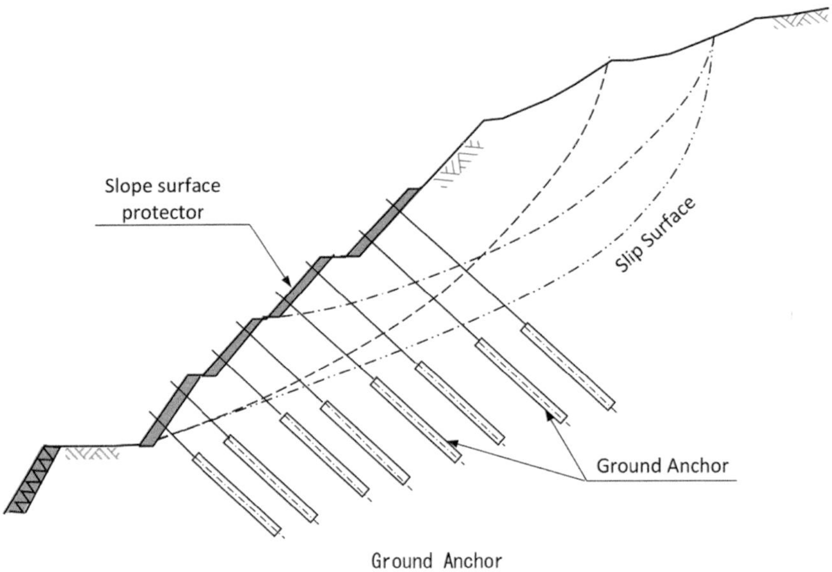

FIGURE 4.10.2
Ground anchors are installed into stable soil under the expected slippage plane.

4.10.3 Reinforcing Steel Bar

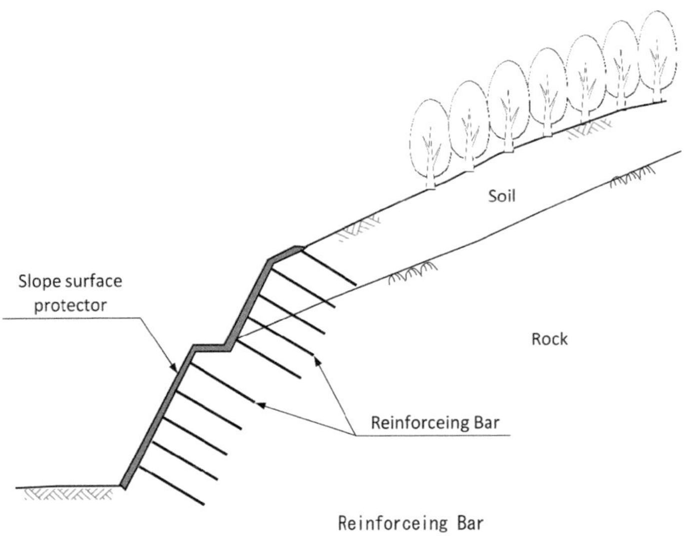

FIGURE 4.10.3
Reinforcing steel bars are inserted into the firm soil. The length, the interval, and the length are designed, considering the external force due to expected slippage.

Afterword

On February 6, 2023, during the writing of this book, an earthquake of magnitude 7.6 struck near the border of Turkey and Syria, causing the collapse of over 220,000 buildings and homes and resulting in the loss of more than 56,000 lives. Furthermore, on the first day of 2024, a damaging earthquake hit the Noto Peninsula area along the Japan Sea coast of the mainland of Japan, and more than 300 people lost their lives.

Besides earthquakes, disasters such as heavy rains, floods, droughts, and forest fires, believed to be related to global warming, are occurring worldwide. In September 2023, a massive flood in Libya, North Africa, resulted in a disaster with over 12,000 dead or missing persons. Additionally, a large-scale forest fire in Maui, Hawaii, and California, USA, claimed many lives.

Since graduating from university, I have been involved in research, development, and practical works for reducing earthquake disasters for over half a century. I have visited disaster areas around the world whenever damaging earthquakes occur. I have repeatedly witnessed the scene of many residents losing their lives buried under the rubble of collapsed houses. Each time, as a researcher in the field of earthquake engineering, I was caught in the regret that "the same thing has been repeated again."

The primary cause of loss of life due to earthquakes is the collapse of residences and buildings with low seismic strength, such as brick structures. The main cause of fatalities due to wind and water disasters is the delay in developing disaster prevention infrastructures, such as river embankments and tidal barriers. The underlying issue is the inability to invest sufficient funds in disaster mitigation due to economic situations.

Domestically, I have experienced two major failures as an expert in earthquake disaster prevention: the 1995 Kobe earthquake and the 2011 East Japan earthquake. The Kobe earthquake made us realize the destructive power of intense seismic motions occurring near inland active faults. The East Japan earthquake resulted in an unprecedented disaster, including the meltdown of nuclear reactors, due to an incorrect assumption about tsunami heights for the design.

Proud of being from a country known for its advanced disaster prevention strategies and technologies, we experts had our pride shattered by these two earthquakes, along with the loss of trust of the people in science and technology. Despite these significant failures, I have spent half a century researching and developing seismic design methods and seismic reinforcement, striving toward building a safe and secure society, albeit slowly. With limited time remaining, I wish to continue this journey without stopping, always moving forward.

I have tackled various challenges such as seismic design for inground structures, lateral movement of liquefied grounds and their countermeasures, safety measures through the filling of inground cavities, and strengthening coastal industrial facilities. These research outcomes will have some effect on the mitigation of future earthquakes and disasters.

This book contains the author and colleagues' challenges and difficulties encountered in the research and practice of earthquake disaster mitigation. If this book can provide some suggestions to young researchers and students in the field of disaster prevention, I would be gratified.

Finally, I express my highest appreciation to Ms. Takako Hamada, daughter-in-law, for her great support in the total compilation of the book, as well as Rintaro Hamada, my grandson, for his excellent translation of the Japanese draft into English.

Furthermore, I appreciate all of the colleagues at the Institute for Disaster Mitigation of Industrial Complexes and its related organizations for their support and cooperation.

I sincerely dedicate this book to Noriko, my wife, for her over half-century devotion to my life as a researcher in earthquake engineering.

October 2025, Yokohama, Japan
Masanori Hamada

References

1. M. Hamada, K. Wakamatsu, S. Yasuda, Liquefaction Induced Ground Deformations During the 1923 Kanto Earthquake, Case Studies of Liquefaction and Lifeline Performance during Past Earthquakes, Technical Report, National Center for Earthquake Engineering Research (NCEER), U.S.A., Volume 1 Japanese Case Studies, 1992.
2. M. Hamada, *Engineering for Earthquake Disaster Mitigation, Springer*, 2013.
3. M. Hamada, *Lessons Learned from 1995 Hyogoken-nanbu (Kobe) Earthquake, Civil Engineering*, Japan Society of Civil Engineers, 1997.
4. M. Hamada, How Were the Lessons Learned from 1995 Kobe Earthquake, *5th National Conference on Earthquake Engineering, Kobe*, 2003.
5. M. Hamada, Lessons Learned from the Great East Japan Disaster and Measures against Future Earthquake, Press in Engineering, 2013, *Proceedings of 4th IDA International Workshop*, 2012.
6. Nuclear Regulation Authority of Japan website. https://www.nra.go.jp/
7. M. Hamada, R. Isoyama, K. Wakamatsu, The 1995 Hyogo-ken (Kobe) Earthquake Liquefaction, Ground Displacement and Soil Condition in Hanshin Area. Association for Development of Earthquake Prediction, Japan, 1995.
8. T.D. O'Rourke, M. Hamada, *Proceedings of the 1st to 8th U.S.-Japan Workshops on Earthquake Resistant Design of Lifeline Facilities and Countermeasures against Liquefaction*, Multidisciplinary Center for Earthquake Engineering Research, (MCEER) U.S.A.
9. M. Hamada, S. Yasuda, R. Isoyama, Permanent Ground Displacement Induced by Soil Liquefaction during 1983 Nihonkai-chubu and the 1964 Niigata Earthquakes, *5th Canadian Conference, Earthquake Engineering*, Toronto, 1987.
10. M. Hamada, S. Yasuda, R. Isoyama, Liquefaction-Induced Permanent Ground Displacement during *Earthquakes, Pacific Conference on Earthquake Engineering*, Christchurch, 1987.
11. M. Hamada, K. Wakamatsu, Liquefaction-Induced Ground Displacements Triggered by Quay Wall Movement, *Soils & Foundations*, Vol.38, 1998.
12. K. Shimamura, Y. Fujita, S. Kojima, Y. Taji, M. Hamada, Transverse Horizontal Load on Buried Pipes Due to Liquefaction-Induced Permanent Ground Displacement, *Soils & Foundations*. Vol.43, No.1, 2003.
13. M. Hamada, T.D. O'Rourke, Large Ground Deformations and Their Effects on Lifelines: 1964 Niigata Earthquake, Case Studies of Liquefaction and Lifeline Performance during Past Earthquakes, Volume 1, Japanese Case Studies, Technical Report NCEER 92-001, 1992.
14. M. Hamada, et al., *The 1999 Kacaeli Earthquake, Turkey, Investigation into Damage to Civil Engineering Structures*, Japan Society of Civil Engineers, 1999.
15. A. Sakamoto, M. Hamada, Liquefaction-Induced Ground Displacement in Uchinada Town by the 2024 Noto Peninsula Earthquake, *Journal of Japan Association for Earthquake Engineering*, Vol.26, No.1, 2026.

16. K. Wakamatsu, N. Yoshida, N. Suzuki, T. Tazoh, *Liquefaction-Induced Large Ground Deformations and their Effects on Lifelines During the 1990 Luzon, Philippines Earthquake Soil and Foundation*, 1991 (in Japanese).
17. M. Hamada, *Critical Urban Infrastructure Handbook*, CRC, 2014.
18. M. Hamada, *Engineering for Earthquake Disaster Mitigation, Earthquake Resistant Design against Long Period Ground Motion*, p. 89, Springer, 2013.
19. M. Hamada, Y. Shiba, O. Ishida, Earthquake Observation on Two Submerged Tunnels at Tokyo Port, *Proceedings of Soil Dynamics & Earthquake Engineering Conference*, Tokyo, 1982.
20. M. Hamada, F. Yonekawa, T. Takeuchi, I. Kato, Reduction of Liquefaction-Induced Ground Displacement by Skip-Driven Pile Method, *Journal of Japan Association for Earthquake Engineering*, 2025.
21. O. Aydan, H. Tano, M. Genis, I. Sakamoto, M. Hamada, S. Yoshimura, Environmental and Rock Mechanics Investigations for the Restoration of the Tomb of Amenophis, *Japan-Egypt Joint Symposium*, Tokyo, 2008.

Index

For Product Safety Concerns and Information please contact our
EU representative GPSR@taylorandfrancis.com Taylor & Francis
Verlag GmbH, Kaufingerstraße 24, 80331 München, Germany